THE POLITICS OF DATING APPS

The Information Society Series
Laura DeNardis and Michael Zimmer, Series Editors

THE POLITICS OF DATING APPS

GENDER, SEXUALITY, AND EMERGENT PUBLICS IN URBAN CHINA

LIK SAM CHAN

THE MIT PRESS   CAMBRIDGE, MASSACHUSETTS   LONDON, ENGLAND

The open access edition of this book was made possible by generous funding from Arcadia – a charitable fund of Lisbet Rausing and Peter Baldwin.

ARCADIA
A charitable fund of Lisbet Rausing and Peter Baldwin

This book was set in ITC Stone and Avenir by New Best-set Typesetters Ltd.

Library of Congress Cataloging-in-Publication Data

Names: Chan, Lik Sam, author.
Title: The politics of dating apps : gender, sexuality, and emergent publics in urban China / Lik Sam Chan.
Description: Cambridge, Massachusetts : The MIT Press, [2021] | Series: The information society series | Includes bibliographical references and index. | Summary: "How dating apps are empowering women and sexual minorities in China, even as they reveal and reproduce systematic sexism and heteronormativity"—Provided by publisher.
Identifiers: LCCN 2020022996 | ISBN 9780262542340 (paperback)
Subjects: LCSH: Online dating—China. | Mobile apps—Social aspects—China. | Sex—China. | Femininity—China. | Sexual minorities—China.
Classification: LCC HQ801.82 .C48 2021 | DDC 306.730285—dc23
LC record available at https://lccn.loc.gov/2020022996

CONTENTS

ACKNOWLEDGMENTS

My interest in dating apps started when I was an active user in the early 2010s. Back then, using dating apps was considered taboo. My friends often expressed serious concern ("Aren't you scared of being murdered?") or pity ("Oh, dear, you don't need to use that to look for a partner") and sometimes even ridiculed me ("You must be randy all the time!"). Almost a decade has now passed, and the situation has not really improved. Thus, researching and writing this book have perhaps been my way of justifying the social legitimacy of dating apps and users like myself.

I first conceived of this book in 2015 as my doctoral dissertation project when I was studying at the University of Southern California (USC). My dissertation advisor, Larry Gross, offered me the most generous and unconditional support from the very early stages of the project. I am extremely grateful to have learned from the wonderful faculty members at USC, including Mike Annany, Manual Castells, Christina Dunbar-Hester, Ange-Marie Hancock Alfaro, Katie Hasson, Yu Hong, Margaret McLaughlin, Patricia Riley, and Karen Tongson, who provided insights and challenged my thoughts in relation to this research. Outside USC, I also want to thank Cassidy Elija, John Erni, Jessa Lingel, Yuxin Pei, Aswin Punathambekar, and Cara Wallis, who on various occasions and in different capacities have given me invaluable advice on researching, writing, and publishing.

I am deeply indebted to the following institutes for their research funding support, without which I would not have been able to conduct the fieldwork for this book in China: the Annenberg School for Communication and Journalism, the Graduate School, and the Center for Feminist Research at the University of Southern California; the Annenberg School for Communication at the University of Pennsylvania; and the School of Communication and Design at Sun Yat-sen University.

I owe special thanks to Yunxi Pei from Sun Yat-sen University and the staff of both Tongcheng 同城 (Gay and Lesbian Campus Association in China) and Qinyouhui 亲友会 (Parents, Families, and Friends of Lesbians and Gays of China) in Guangzhou, who helped me recruit research informants. Xuan Chen, a graduate student I met at Sun Yat-sen University, also volunteered to assist with my fieldwork.

Writing and revising are long and sometimes lonely processes. Half of this book was written as my doctoral dissertation. I must thank the dissertation writing group members at USC—Cat Duffy, Michelle Forelle, James Lee, Nathalie Maréchal, Raffi Sarkissian, and Sarah Myers West—for their company on my writing journey. My other friends at USC, including David Jeong, Minwoo Jung, Melina Sherman, and Andrea Wenzel, also listened to my thoughts and provided feedback on earlier drafts of some chapters of this book.

I want to thank Kei Imafuku for his patience in listening to my repeated venting and nascent ideas. I am blessed to have my parents and my brother, who have always been supportive of my academic endeavors, regardless of which continent they have taken me to.

Credit goes to Andrew Schrock, the anonymous reviewers of this book, and Gita Manaktala and her team at the MIT Press. I am grateful for their encouragement and extremely constructive feedback at various stages of my writing.

Finally, I owe my sincere gratitude to all my informants, who were generous in sharing their stories with me and promoting my research to their friends and acquaintances. Each of them has a unique story, which I regret that I cannot fully cover in this book.

An earlier version of chapter 2 appears as "Liberating or Disciplining? A Technofeminist Analysis of the Use of Dating Apps among Women in

Urban China" in *Communication, Culture & Critique, 11*(2) (2018), published by Oxford University Press. Chapters 3 and 6 include some parts of my article "Multiple Uses and Anti-Purposefulness on Momo, a Chinese Dating/Social App," *Information, Communication & Society, 23*(10) (2020), published by Taylor and Francis. I thank these publishers for their permission to reprint the articles.

NOTES ON TRANSLATION AND TRANSLITERATION

As this book is about the Chinese dating app culture, I include dozens of local terms. For every term, in its first appearance in each chapter, I offer its Chinese characters, *hanyu pinyin* form 汉语拼音, and English translation.

I translated all of the Chinese interview extracts into English. I also translated the other Chinese reference materials, such as news articles and dating app profiles, into English.

Chinese put their family names before their given names. I follow this conventional order when I refer to historical and public figures. When I cite a Chinese academic, I use the Western order: given name followed by family name.

1

INTRODUCTION: DATING APPS HAVE POLITICS, TOO

Two months ago, I saw two of my female friends swiping with their fingers on their phones and uploading their selfies. I thought it was very interesting because on apps like QQ and WeChat many people do not use their real photos. I am not interested in people who do not post real photos because you cannot tell what they look like or how old they are. But on this app, the photos look real, and people mention their hobbies and interests. My female friends told me that their friend had found a boyfriend through this app. Well, I thought, it is fun to upload photos, so I downloaded the app.

This was Nancy recalling why she first downloaded Tantan—a major dating app in China—onto her mobile phone.[1] When we first met in October 2016, mobile dating apps have already become a global phenomenon. In English-speaking countries, Tinder and Grindr are popular dating apps for straight users and queer male users, respectively. However, China has its own dating app ecology, the same as its other internet services. Momo, founded in 2011, is now the most widely used dating app among heterosexual Chinese. By the end of 2017, it had nearly 100 million active monthly users ("Guanyu Momo," 2019). Tantan, launched in 2014, was reported in early 2018 to have over 20 million active monthly users ("A Guy Who Grew Up in Stockholm's Suburbs," 2018).[2] For queer men, Blued and Aloha are the market leaders, while Rela, Lesdo, and Lespark are popular among queer women in China.

In this book, I argue that these apps—with their unique market positioning, color tone, and interface design—are not merely software running on mobile phones. They are portals that transport people from their mundane physical environment to an exciting virtual world full of relational possibilities. At the same time, dating apps are sites of power dynamics, impacted by the heteronormativity and patriarchy infused throughout Chinese society.

Admittedly, Chinese dating apps are a challenge to describe as a coherent genre because companies are constantly expanding and diversifying their features. However, there are five main features shared by most Chinese apps that, as I later argue, have specific affordances that constitute the "networked sexual publics" I explore in this book. The first commonly found feature is the "people nearby" feature, resembling the classic Grindr interface. An array of profiles is arranged in ascending order of distance using geolocation data, and any user can send messages to people nearby. The second feature is the "swiping" feature pioneered by Tinder. Users are presented with a series of photographs of others, one at a time. If they are interested in someone whose photograph pops up on their screen, they swipe right across that person's photograph to signal interest. Users can talk to each other only if both parties express an interest in each other. In Western apps, the "people nearby" feature and the "swiping" feature are generally mutually exclusive; however, some Chinese apps offer both. Third, there is the "groups nearby" feature where groups of various topics are created by users. This is akin to web-based forums, the only difference being that the members of the groups are physically nearby. The fourth feature is similar to the "status updates" function of Facebook that broadcasts a message or picture to network connections or people nearby. Finally, the "live streaming" feature appears in most Chinese dating apps. Users can sign up as a live streamer and broadcast live video content to anyone.[3] This feature is a profit center for apps such as Momo and Blued (Deng, 2018; Edmunds, 2017; S. Wang, 2019a, 2019b). Viewers can purchase digital gifts from the apps and send these gifts to their favorite live streamers. The apps take a share of the gifts' monetary value and distribute the remainder to the streamers.[4]

Single and in her late twenties, having had only two previous romantic relationships, Nancy was extremely enthusiastic to find her Mr. Right. She said,

> On WeChat, there is the "people nearby" feature that I can use to meet someone that I don't know. But many people do not put up their photos. . . . I prefer using Tantan to make friends. Although many Tantan users are not there for friends, but for, you know, I can at least upload my photos there and get a few "likes." *(Researcher: How many matches do you have so far?)* More than a hundred.

The millions of Chinese like Nancy who use these apps have specific ways of referring to them. In China, these apps are known as *jiaoyou chengshi* 交友程式 (friend-making apps) or *jiaoyou ruanjian* 交友软件 (friend-making software). The term *yuehui chengshi* 约会程式, which is a direct translation of "dating apps," is not often used. *Yuehui* 约会 consists of the words *yue* (to arrange; to make a reservation) and *hui* (meeting) and therefore refers to a date. For readers who are familiar with the Chinese dating app culture, the term *yuepao shenqi* 约炮神器 will be most familiar. *Yuepao* 约炮, which consists of the words *yue* and *pao* (cannon), is a neologism for "hooking up," using *pao* to hint at the phallus. On apps such as Momo and Blued, users might send a one-word message, *yue?* 约? (meet?), to solicit hookups.[5] *Shenqi* 神器, consisting of *shen* (god) and *qi* (tool), stands for "a powerful tool" or "a magical tool." Therefore, the phrase *yuepao shenqi* means "a magical tool for hookups." The phrase's origin is hard to date, but it was popularized in a viral video made in 2012 by Michael Stephen Kai Sui, an American actor who speaks fluent Chinese.[6]

Researchers conducting some of the earliest studies alternately referred to these apps as "social apps," "hookup apps," "location-aware dating apps," and "geosocial networking apps" (e.g., Albury & Byron, 2016; Blackwell, Birnholtz, & Abbott, 2015; T. Liu, 2016; Rice et al., 2012). For simplicity, in this book I use the linguistic shorthand *dating apps* to refer to this genre of communication technology. However, as I describe later, these apps do not market themselves solely as dating apps, nor do the users of these apps use them only for locating romantic partners. For these reasons, although I employ the term *dating apps* to refer to a shared genre of mobile software, this book has a more expansive perspective on how users interpret and use these apps for more than just dating. Returning to my conversation with Nancy, she had no intention to *yuepao* on Tantan. In fact, she thought the app could protect her from unsolicited sexual requests and allow her to look at handsome men. She said,

> Because [Tantan] has a list of preselected keywords, for example, *yuepao*, the system will remind you to report [harassment]. That is to say, I think the original intention of this app's design is good; it is not meant to make chaos. There are still regulations [from the administrator]. . . . I am a person who judges others based on their face. Because if you don't understand someone's mind, you definitely start with their appearance. If their appearance is okay, then I will want to know more about them. . . . When my female friends looked at the app, they kept saying, "He is so handsome." I thought, "Wow, it is true. There are many good-looking men on this app."

The way Nancy described Tantan implies that the app has empowered her by allowing her to report sexual harassment. From Nancy's perspective, Tantan reverses the conventional pattern of male gaze where men are the subject and women are the object (Mulvey, 1975). She could look at and learn more about potential partners without the constant burden of sexual requests.

I met Nancy again in September 2018 for a cup of coffee. Having communicated with her on and off via WeChat for around two years since we first met, I decided to disclose my sexual orientation to her. I casually mentioned making a plan with my boyfriend. On hearing about my boyfriend, she swallowed her food, looked me in the eye, and said, "I think I like women, too." I was shocked by her coming out because in our first interview in 2016 she identified as straight. Further, after the first interview, she told me she had a date with a man she had met on Tantan, and sometimes she asked me how dating in the United States differed from that in China. I had never expected her to come out to me as a queer woman. She said that she recently started questioning her sexuality because she had not been interested in men for a year. She also said that because there was a very visible LGBTQ culture in her hometown, she was open to the idea that she might be a bisexual woman. We then talked about the popular apps queer women in China were using, and I encouraged her to try some of these to experiment with her sexuality.

Nancy's experience is one of the many stories this book documents and analyzes. Dating apps are a new platform for personal relationships, and it is worthwhile to examine how relationships are shaped and recon-figured by this emerging technology. However, this book takes a critical perspective to expand our understanding of the opportunities and chal-lenges dating apps have presented for their users. What does a "dating app" mean for a person like Nancy, who is coming to understand herself as a bisexual woman? How do these apps help their users, straight or queer, female or male, cohere new forms of publics?

In this introductory chapter, I lay an epistemological, theoretical, and methodological foundation for the book. First, I argue for a need to exam-ine dating apps from a critical perspective. To justify this perspective, I introduce three sets of interrelated literature that have informed my research and thinking. I also provide a brief account of my research pro-cess and highlight some key ethical issues in researching sensitive topics like sexuality (a detailed reflection is included in the appendix). Finally, I provide an overview of the chapters in this book, describing how a criti-cal perspective helps us understand dating apps in the context of straight and queer Chinese.

TWO TRADITIONS OF DATING APP RESEARCH

Dating app studies have burgeoned since the early 2010s. Drawing on theories and approaches from the social scientific tradition, researchers have explored various issues related to interpersonal processes, such as motivations, uses, self-presentation, relational development, and risks.[7] These lines of inquiry are paramount. People do develop interpersonal relationships, whether romantic or sexual, long-term or short-term, involved or casual. In my earlier work, I called the form of intimacy that is facilitated by dating apps "networked intimacy" (L. S. Chan, 2018a). My concept is similar to "mobile intimacy" (Hjorth & Arnold, 2013), "mediated intimacy" (Attwood, Hakim, & Winch, 2017), and "virtual intimacy" (McGlotten, 2013) but highlights the networked nature of the extensive connectivity supported by dating apps.[8] Networked intimacy, as I have argued, has a built-in ambivalence where the relational goals people state on their dating app's profile may not always correspond to what they actually want or eventually get, a well-curated profile is nec-essary but may not be taken seriously, and extensive connectivity to a large pool of potential partners can be both a blessing and a curse. The everyday personal use of dating apps indeed provides fruitful material for social scientific research.

However, since that point I have become increasingly concerned with the relationships between dating apps and power in the context of Chinese society. As second-wave feminism reminds us, "the personal is political."[9] That is, there is always a political dimension underlying personal relationships, where personal issues are never insulated from powerful social forces, and vice versa.[10] By *political*, I do not mean the narrow sense of electoral politics or civic engagement. Rather, I mean the constant struggle for and distribution and maintenance of power in society. Politics, therefore, is a power relationship. Taking this broader conceptualization of politics—as Kate Millett did in her groundbreaking work *Sexual Politics* (1978)—I use the phrase *gender and queer politics* to refer to how traditional gender roles and heteronormativity entangled in patriarchalism have contributed to the unequal treatment of women and queer people in society. This critical tradition is what this book is built on.

This view of power is consistent with Michel Foucault's (1978) observation that "relations of power are not in a position of exteriority with respect to other types of relationships (economic processes, knowledge relationships, sexual relations), but are immanent in the latter" (p. 94). Drawing on Foucault, I consider intimacy as just such a site of power struggles. For this reason, throughout this book I regard dating apps not just as technological artifacts but also as a lens to reveal power relationships. As Langdon Winner (1980) notes, technological artifacts have politics. Some technologies create order in our world that favors certain people while disadvantaging others. Other technologies require a specific set of social conditions for their proper operation. Whether creating new social relationships or reproducing existing cultural conventions, technologies do political work in society.

This book extends the insight that technologies have politics to help us understand the politics of dating apps. Dating apps may create an order that favors users of a certain gender and sexual orientation and discriminates against others. They may fit into an existing social order that is already biased against some people, or they may create a new order that resists the existing one. Accordingly, I also regard dating apps as what Ara Wilson (2016) calls "the infrastructure of intimacy." For Wilson, "infrastructure offers a useful category for illuminating how intimate relations are shaped by, and shape, materializations of power" (p. 263). For example, she argues that the design of public bathrooms is predicated on gender segregation, while telecommunications technologies such as the telephone break down spatial divisions. By conceiving dating apps as infrastructure, I shift my attention from a pure concern with personal relationships—the typical social science perspective on dating apps—to the negotiations, opposition, and subversion of power in intimate relationships.

NETWORKED SEXUAL PUBLICS

To capture the political dimension of the emerging dating app culture, I propose a new concept of networked sexual publics. *Publics* has always been a broad, unifying concept. Sonia Livingstone (2005) defines *publics* as a collection of people sharing "a common understanding of the world, a shared identity, a claim to inclusiveness, a consensus regarding the collective interest" (p. 9). Because the word *publics* does not imply essentializing or homogenizing attributes, it can capture collective relationships that people imagine to be real. Larissa Hjorth and Michael Arnold (2013) point out that a public can be the "phenomenological reality" that a person can see, smell, hear, and touch. At the same time, publics are driven by the "existential imaginary" because a member can often only imagine someone else is sharing a common experience with them. In the context of this book, I believe gender dynamics and queer possibilities are driven by these mutually reinforcing tangible and imaginary qualities of publics.

While publics have always existed, they are increasingly connected by the global media and social networking sites. Scholars argue that networked publics came into being in the late twenty-first century with this technological shift. Mizuko Ito (2008) explains that the term "references a linked set of social, cultural, and technological developments that have accompanied the growing engagement with digitally networked media" (p. 2). Ito emphasizes that the notion of *publics* implies more engagement than *consumer* or *audience*. danah boyd (2011) further develops the term *networked publics* to describe social groups driven by the rise of networked technologies such as Facebook. The affordances of social media—including persistence, replicability, scalability, and searchability—have reconfigured public engagement, allowing members of the public to "gather for social, cultural, and civic purposes" and "connect with a world beyond their close friends and family" (p. 39).[11] I will shortly expand on the relationship of affordances to dating apps and the publics they bring into being. For now, I am merely drawing attention to how boyd uses the term *networked publics* to refer to both digital space and the collection of people it connects, expanding our understanding of publics.

However, some scholars argue that the metaphor of networks is inadequate in capturing the form of sociality facilitated by mobile media. Larissa Hjorth and Michael Arnold (2013) contend that a networked metaphor "privileges ramified dyadic relationships [and] fails to signify collectivity, emotional affect and a shared horizon" (p. 12). They further elaborate:

> The network is also transient and shape-changing. . . . The central node may know the network, and nodes know connecting nodes, but the network is too dynamic and ephemeral to be present to itself as a common entity. There is therefore no sense of solidarity across the network, no sense of tradition, no common identity (in contrast to individual identities) and no common interests (in contrast to individual interests). (p. 133)

This image of network as painted by Hjorth and Arnold is partly true for dating apps. If we empirically analyze dating app cultures, we know that users use dating apps mainly to look for dyadic relationships. New users join and old users quit these apps all the time, which makes the networks of dating apps inherently transient. As a user and as the node of your own network, you know who you have connected to but may have no idea whether a person has spoken to another person. However, in contrast to Hjorth and Arnold, I believe that dating apps can provide a common identity or solidarity. In this book, I argue that because dating apps in China in general are built for a specific community—either straight men and women, queer men, or queer women—their existence is predicated on a common identity and shared history. To me, whether or not solidarity can be forged on dating apps is an empirical question rather than a conceptual debate. Therefore, I invoke the term *networked* in the phrase *networked sexual publics* to highlight the primary architecture of dating apps, which is a web of dyadic connections that create and sustain intimacy. These connections are, in the words of Hjorth and Arnold, the "phenomenological reality" of most users.

There are conceptual similarities between my concept of networked sexual publics and the concept of digital intimate publics as formulated by Amy Shields Dobson, Brady Robards, and Nicolas Carah (2018). Commenting on a much broader phenomenon of public-facing lives on social media, Dobson and her colleagues think about "digital intimacy" as both social capital and labor. On the one hand, digital intimacy is about connections and relations that may convert into resources and mobility, acting as social capital. On the other hand, digital intimacy also builds various types of relationships that, through algorithms, produce data for social media companies, a process that requires labor. Likewise, dating apps, as I document in this book, allow users to make friends, locate partners, find business collaborators, and search for their communities. Using dating apps also requires crafting a profile, deciphering others' messages, and handling malicious or negative responses. Shuaishuai Wang (2019a, 2019b) examined how dating app companies (in his case, Blued) monetize the performative labor of their live streamers. As with other conduits for digital intimacy, social capital and labor are built into dating apps. For this reason, as I discuss shortly, I treat power as a central concern of networked sexual publics.

The term *sexual publics* refers to a particular subset of publics. As Katherine Sender (2017) puts it, "sexual publics describes a loose affiliation of members who, for whatever brief or extended period, see themselves as part of a shared experience of mediated sexuality" (p. 75). In a space like a museum, even though viewers do not occupy the same space at the same time, being inside a museum allows viewers to imagine the existence of other viewers in the past and in the future. This points to the "existential imaginary" of the publics Hjorth and Arnold (2013) have discussed as an aspect that network-only definitions of *publics* neglect. In Sender's analysis, the imagination is forged by the shared space of a sex museum. In this book, dating apps act as a foil for the imagination.

These insights help bolster my concept of *networked sexual publics* to refer both to the technological network of people like Nancy, who are united by their shared position in the patriarchal and heteronormative world and connected by dating app technologies, *and* the space where a multiplicity of interpretations and relationships for the publics is possible. From this perspective, Nancy's experiences with dating apps are no longer banal. Selecting a romantic partner, looking at handsome men, acquiring social recognition by accumulating "likes," reporting men's sexual harassment, and possibly experimenting with her sexual orientation are important ways she grapples with her place in the networked sexual publics. Networked sexual publics are where women and men negotiate their power and sexual minorities strive for a place in the heteronormative world. They expose and reproduce systematic sexism and heteronormativity. Simultaneously, they create a digital environment for negotiation, subversion, and potentially backlash from, as I show in this book, straight men and the queer community itself. By looking at the experiences of dating app users, including female and male, straight and queer, this book documents and analyzes the struggles of networked sexual publics in China. In so doing, it broadly examines how dating apps serve as a tool for empowering women and queer people in the country.[12]

In the next few sections, I situate my research framework at the intersection of three theoretical traditions. The first body of literature examines how technology has always played in a role in shaping gender relations and queer lives. The second explains why I foreground users' agency in their use of technology. Finally, I contextualize my study in the emerging scholarship on mobile cultures in Asia-Pacific.

TECHNOLOGY, GENDER, AND QUEER

Feminist scholars have long been concerned with the relationship between technology and gender politics. When Wendy Faulkner (2001) traces the development of technological concerns in feminism, she identifies three streams. First, what Faulkner calls "women in technology" concerns the inclusion of women in the technology industry through their earlier socialization into "machines" and the changes to workplace policies (such as setting up childcare). These measures have been advocated by liberal feminists. The second stream is "women and technology," which focuses on the use or the receiving end of technologies. Faulkner sees scholars in this stream examining the effect of technologies on women. For example, the proponents of cyberfeminism have been very optimistic about the liberating potential of the internet. They have been intrigued by how users can take up alternative identities, thereby breaking gender binaries and transforming conventional gender roles (Plant, 1997). However, these two streams of scholarship have often failed to consider the symbolic association between masculinity and technology, thereby neglecting one side of gender dynamics.

The third stream, "gender and technology," acknowledges the mutually constitutive nature of gender and technology. A particular framework from this stream is technofeminism, where "gender relations can be thought of as materialized in technology, and gendered identities and discourses as produced simultaneously with technologies" (Wajcman, 2007, p. 293). Instead of just considering women's issues, some scholars in this stream have concomitantly examined how technology and masculinities intertwine. For example, Ellen van Oost (2003) documents the development of Philips's razors. Philips's first razor in 1939 targeted both men and women. However, in the 1950s, the second generation of razors began differentiating between male and female users. Men were seen as technologically competent; therefore, the designs of the shavers revealed their internal technology and came with a lot of information and controls. In contrast, women were conceived of as technophobic. For this reason, female shavers were marketed not as electronic appliances but as cosmetic products. They came with one simple button and were assembled using a "click" rather than screws, which connote technology. In this case, the technology "not only reflected this gendering of technological competence, they too constructed and strengthened the prevailing gendering of technological competence" (p. 207). Analyzing the marketing of Philips's razors gave insight into how a specific technology became gendered.

Communication technologies have had a close relationship with queer possibilities and heteronormativity. Although the academic enterprise of queer science and technology studies has only gradually been formulated in the last decade or so (Landström, 2007; Molldrem & Thakor, 2017), the significant research and commentary found in communication and media studies since the 1990s have examined how the internet serves as a liberating space for queer people. Cyberspace, as imagined by science fiction writer William Gibson in his 1982 short story "Burning Chrome" (T. Jones, 2011), was a place where ideas can freely compete and flow. Queer people similarly use the internet as a cyberspace to share and access content because of its relatively low cost compared to traditional media. In particular, teenagers who are isolated socially and financially from urban queer culture use the internet to explore their sexuality (Campbell, 2004; Cassidy, 2018; Gross & Woods, 1999; Mowlabocus, 2010). Mary Gray (2009) documents how websites have become crucial ways queer youth living in rural America discover their sexual desires and search for belonging. As one of Gray's informants said, "If I didn't have access to computers, I don't know what I would do" (p. 137). Andre Cavalcante (2019) argues that the nonexistence of a real-name policy on Tumblr and its reblog feature have allowed queer youth to explore their identity with privacy and anonymity. Accordingly, the internet has contributed to what Lauren Berlant and Michael Warner (1998) call "queer world-making," which is to transcend the "the logics of compulsory heteronor-mativities" (West, Frischherz, Panther, & Brophy, 2013, p. 56). In each of these examples, networked technologies have enabled queer youth to find others and build community.

For these reasons, the rise of dating apps in the last decade has radi-cally changed how queer people connect with each other. The location awareness and the dominance of visuals of these apps have promoted physical encounters and embodiment. Dating apps' subversive potential reside in their ability to facilitate same-sex intimacies, which are not com-pletely socially acceptable throughout the world or are, in many coun-tries, still illegal. And although scholars have agreed that the internet has helped queer people form or seek their communities, they have been less certain of whether dating apps such as Grindr—which is often reduced to facilitating sexual encounters—can do the same. On dating apps, users meet strangers who are physically nearby but mostly exist as individuals. Sam Miles (2018) points out that "the potential for these strangers to become social or sexual partners, and in turn more significantly represen-tative of community . . . remains under-theorized" (p. 8). By infusing the literature on dating with concepts about publics and politics, I counter this tendency to regard every encounter on dating apps as simply related to interpersonal communication leading to sex.

The liberating potential of the internet and dating apps has also created moral panic. Since the 1990s, governments have begun controlling the circulation of sexual images on the internet under the banner of "protecting the children" (Gross & Woods, 1999). This rhetoric has remained strong in the 2010s, exemplified by Tumblr's banning of all adult content in December 2018, claiming it had "a responsibility to consider that impact across different age groups, demographics, cultures, and mindsets" (D'Onofrio, 2018, para. 5). Commentators argue that this ban is detrimental to the LGBTQ communities (Reynolds, 2018). In addition, dating apps have been blamed for the spread of sexually transmitted diseases both in China and Western countries (Brait, 2015; "Chinese Gay Dating App Blued Halts Registration," 2019; Parry, 2015). This accusation, because it was not welcomed by app companies, has prompted the industry to integrate safe-sex promotional messages and testing notifications into their apps (Kraus, 2018). Examining queer politics requires the simultaneous consideration of users' practices, state regulations, and commercial practices.

Scholarship above has shown that technology, in various ways, shapes gender and queer politics. However, I have presented two separate threads of research—one engaging with technology and gender and another dealing with technology and queerness. While there is a burgeoning subfield of queer feminist science studies (see Cipolla, Gupta, Rubin, & Willey, 2017), there is a dearth of theoretical orientations and empirical studies that simultaneously address gender *and* sexuality in relation to technologies such as digital media. This should not be a surprise because scholars often orient themselves to either feminist studies or queer studies, which has resulted in an inescapable tension. When tracing a genealogy of queer feminism, Mimi Marinucci (2010) observes a history of tension between the two. In canonical feminist literature, lesbianism, gay male drag performance, and transgenderism are criticized or neglected. Similarly, women's experiences are often downplayed in queer studies. The abyss between feminist and queer scholarships is also widened when my field of communication departmentalizes subject interests. For instance, the International Communication Association has created a distinction between the Feminist Scholarship division and the LGBTQ Studies Interest Group. A lack of a coherent theoretical framework that unifies the discussion across gender and queer issues in relation to dating apps or digital media at large is a natural result of the historical tension Marinucci identifies. Clearly, exploring the role of dating apps, as an emerging communication technology, involves investigating power dynamics related to both gender and sexual orientation. For this reason, I intend my concept of networked sexual publics to be a conscious intervention to bring these worlds together.

INTERPRETATIONS, AFFORDANCES, AND COMMUNICATION TECHNOLOGIES

Among the various theoretical concepts I rely on in this book, the dual concepts of interpretation and affordances are key to understanding the relationships between dating apps and gender and queer politics. The word *interpretation* refers to the way people assign meanings to things and events. For example, a folding chair can be interpreted as a support for one's body weight or as a weapon. I draw my understanding of interpretation from the social construction of technology (SCOT) literature. SCOT is concerned primarily with how social forces drive technological development (Baym, 2010). Treating technological development as a series of variations and selections, an analysis based on SCOT identifies the social groups that are relevant to a technological artifact, like designers, retailers, regulators, and investors (Pinch & Bijker, 1987). Because these groups have distinct types and degrees of political, social, and economic power, they define the artifact's problems differently. The availability of disparate ways to define these problems is referred to as *interpretive flexibility*. Various interpretations compete, and the winning interpretation drives the development of the artifact.

Interpretive flexibility explains how people use a technology in a way that its designer did not intend (Oudshoorn & Pinch, 2003). For example, a child may use a plastic bucket as a helmet. Ronald Kline and Trevor Pinch (1996) show how in the early 1900s, automobiles in rural America were interpreted as engines to drive grinders, saws, and washing machines—not purely for transportation. In their words, "this flexibility was not at the design stage. New meanings are being given to the car by the new emerging social group of users" (p. 777). Car dealers that noticed this unintended use by farmers sold kits that could turn a car into a power source. The automobile manufacturer Ford later banned these kits, demonstrating the unequal power dynamics where companies can exert control on users through the artifact and laws.

The concept of *affordance* moves us closer to a concern with technological agency by simultaneously considering users' interpretations and the materiality of the technology. The term was coined by James Gibson (1979) to describe "the complementarity of the animal and the environment" (p. 127). Affordances are not simply the features of an environment but the interactions between such features and the animals within the environment. When an animal in a field sees a cave, it interprets the cave as a place for shelter, not as a rock-based structure with an empty inner space. The cave affords sheltering for an animal.[13] The concept was successively introduced to the sociology of technology as a "third way between the (constructivist) emphasis on the shaping power of human agency and the (realist) emphasis on the constraining power of technical capacities" (Hutchby, 2001, p. 444). Ian Hutchby agrees that "environments or artefacts have affordances which enable the particular activity while others do not" (p. 448). Yet "these affordances *constrain the ways that they can possibly be 'written' or 'read'*" (p. 447, emphasis in original). That is, affordances both enable and constrain users' actions and interpretations.[14] The affordance of support of a folding chair enables us to recognize and sit on it. The materiality of it also, according to Hutchby, limits our interpretation of the utility of such an object: a folding chair can never be interpreted as a flying machine.

To examine the role of communication technologies in our behaviors and social change, media researchers have also turned to affordances (Baym, 2010; Neff, Jordan, McVeigh-Schultz, & Gillespie, 2012). In the process, some have clarified what constitutes an affordance. For instance, Peter Nagy and Gina Neff (2015) reconceptualize affordances as a combination of (1) the materiality of an object, which refers to the object's features; (2) the mediated experience of the object, which includes the users' expectations and knowledge of how the object can possibly be used; and (3) the emotional states of the users, which influence how they perceive the features of the object and perform actions. This tripartite model foregrounds neither the object nor the users. The second and the third components suggest that affordances are highly individualized because each user can have a unique experience of an object and experience different emotions. Sandra Evans, Katy Pearce, Jessica Vitak, and Jeffrey Treem (2017) argue that affordances are not the features of an object and that unlike features which are either present or absent, affordances exist on a continuum. Further, they suggest that affordances should not be mistaken for the consequences of using an object. They insist that "an affordance can be associated with *multiple* outcomes" (p. 40, emphasis in original).

Communication scholars, including myself, have conceived typologies to describe the affordances of communication technologies. Andrew Schrock (2015) suggests four communicative affordances of mobile media—portability, availability, locatability, and multimediality. Jessica Fox and Bree McEwan (2017) consider ten types of affordances of communication channels—accessibility, bandwidth, social presence, privacy, network association, personalization, persistence, editability, conversation control, and anonymity. In my earlier research, I proposed five affordances of dating apps (L. S. Chan, 2017b):

1. Mobility: With dating apps, users have access to numerous potential partners anywhere and at any time. This is similar to Schrock's (2015) idea of *portability*. This affordance is fundamental to any mobile media that rely on wireless connections.
2. Proximity: The global positioning system built into our smartphones enables users to look for others who are physically nearby. Such system also affords what Schrock (2015) calls *locatability*. People can coordinate with each other or monitor others' location with the assistance of the system. Proximity seems to be particularly important to dating apps because other communication technologies such as mobile phones or videoconferencing place a heavier emphasis on social presence (Fox & McEwan, 2017) than physical presence.
3. Immediacy: Dating app users can meet each other quickly. Immediacy is closely related to *proximity*. In some dating app user communities, users are expected to meet up quickly (Licoppe, Rivière, & Morel, 2016).

4. Authenticity: Using data from other social media platforms, some dating apps verify users' information. Some apps reveal how many friends two users have in common. This affordance, however, negates anonymity (Fox & McEwan, 2017).

5. Visibility: Most dating apps provide a dominant screen space for users' photographs. Users are expected to upload an attractive "profile pic." Other apps allow users to input detailed written information.[15]

Although the concept of affordance has been developed in multiple fields, a core tenet has endured: neither technological features nor interpretations alone determine how the technology can be used and what consequences it may have. Affordance captures this contingency resulting from people's subjective perceptions of a technology based on its objective features. Throughout this book, I emphasize the agency of app users, who actively and creatively make use of the various affordances of dating apps for a variety of practices. Because users occupy different positions in the patriarchal and heteronormative structure, their interpretations of dating apps also differ. In this book, affordance is an idea that illuminates the agency of users in their particular social positions as they use dating apps while retaining my overarching concern with power.

MOBILE CULTURES IN ASIA-PACIFIC

The third and last theoretical tradition that anchors this book comes from the emerging scholarship on mobiles cultures in Asia-Pacific. Interest in lively, multifaceted mobile cultures in non-Western contexts has grown since the early 2000s. For instance, the anthology *Mobile Cultures: New Media in Queer Asia* (Berry, Martin, & Yue, 2003) represents one of the earliest scholarly engagements in this area. The central debate that this collection of essays explores is how the globalization of sexual cultures has contributed to a homogenized or a heterogenized sexual landscape in Asia-Pacific. It is homogenized because the circulation of Western gay culture through mass media, commodity, and tourism has at times permeated into Asian queer communities. It is heterogenized because the very same mechanism that allows Western or American gay culture to enter into Asia has opened up multiple paths where, in the words of

Ann Cvetkovich and Douglas Kellner (1997), "local forces and situations mediate the global, inflecting global forces to diverse ends and conditions and producing unique configurations for thoughts and action in the contemporary world" (p. 2).

This issue is recently taken up in *Mobile Media and Social Intimacies in Asia: Reconfiguring Local Ties and Enacting Global Relationships* (Cabañes & Uy-Tioco, 2020), an edited volume that examines the way ubiquitous mobile technologies have configured intimate relationships and coins the term *glocal intimacies*. The volume contends that globalization is not a one-way homogenization and that strong local sociocultural power dynamics remain. When writing this book, I continued responding to this still-unfolding challenge.

Although Berry et al.'s *Mobile Cultures* was published nearly two decades ago, there are at least two lessons to be learned from this anthology. First, when conducting research in non-Western contexts, it is crucial to historicize the development of digital media. For example, the emergence of the queer communities in the United States and Western Europe predates the widespread use of the internet. However, in Asia, the chronological order is reversed because queer communities have largely been suppressed socially and politically (Berry & Martin, 2003). This difference suggests that theoretical insights about the role of the internet in the development of queer communities may not be able to translate well across geographical contexts. Second, a thorough understanding of the local language and culture is a prerequisite for nuanced, insightful analysis. For instance, *lazi* 拉子 used in Taiwanese online bulletin board is a transliteration of the English term *lesbian*; it is also the nickname for a character in a local popular novel in the region about lesbian relationships (Berry & Martin, 2003). Therefore, when the term *lazi* is used online, what is articulated is not just an appropriation of Western ideal of lesbians but also a reference to the local queer culture. An outsider not well versed in the cultural meanings of certain phrases and symbols might miss the full meaning of communication.

I place my attention on China in this book because the entire Asia-Pacific region is not monolithic. For example, Larissa Hjorth (2008) investigates the culture surrounding mobile phones in four different Asia-Pacific locations. She proposes the notion of "cartographies of personalization" to depict how women in these four locations have domesticated mobile phones. In Japan, short novels written for mobile phone viewing were dominated by female writers and readers. In South Korea, women gave opinions on what ringtones and wallpapers their male partner should use on their phone as a way to show to the world that their male partner already had a girlfriend. In Hong Kong, Hjorth notes that mobile media provide a platform for nostalgia on an individual level. In Australia, in mobile media she finds "postal presence"—the sense of copresence created by embedding one's photograph in a text message. In this multisited study, mobile phones become a lens through which we can look into the cultural processes underlying everyday life. The growing literature on mobile cultures in Asia-Pacific underscores the need to place mobile technologies in the region's broader environments. At this point, I think it is crucial to contextualize the rise of dating apps in China.

RISE OF DATING APPS IN CHINA

In the specific context of China, I consider dating apps as a "lash-up" (Molotch, 2003) of two analytically distinct but interrelated transformations—one primarily social, another economic. The social refers to the transformation of intimacy. Anthony Giddens (1992) observes the rise of the pure relationship model, replacing the procreation-driven model of intimacy in Western countries. A similar phenomenon has taken place in modern China. Because traditional Chinese culture did not have the concept of *ai* 爱 (love), love was not prioritized in traditional marriage (Cheung, 1999). Instead, marriage was determined by one's parents (Fei, 1939). The founding of modern China in 1949 brought a change in courtship and marriage practices. In cities during the Maoist period, men and women were assigned to work in *danwei* 单位 (work units). Under the planned economy system, the *danwei* was both an economic and a social organization. *Danwei* leadership actively intervened in personal affairs, including matching couples and mediating marital conflicts (J. Liu, 2007). The rarity of switching *danwei* also meant that people had a very limited pool of potential spouses.

   With the decollectivization of *danwei* during China's economic reform, the state stopped organizing matching activities (Zhang & Sun,

2014). Living within these rapidly changing parameters for intimate life, the younger generation in China developed new dating practices. Unlike members of the previous generation, who rarely had more than one dating relationship before marriage, the younger generation saw their dating partners not necessarily as their future wives or husbands (Farrer, 2002). Without the pressure from their *danwei*, queer people could possibly get away from the norm of heterosexual marriage (D. Wong, 2015). Lisa Rofel (2007) suggests that market reform in urban areas during post-Mao China has unleashed desires that were suppressed in the Maoist era, constituting the new "desiring subjects." This social change has provided a context for the popularity of dating platforms.

The economic transformation refers to the economic reform initiated in 1978. With the opening up of the economy and various industries, increasingly more Chinese enterprises have adopted a profit-driven model. There are both a general economic consideration and a specific political economy underlying the development of Chinese dating apps. On the general ground, online dating and marriage-matching services face a challenge in retaining their customers because customers will not return after finding a lifelong partner through their services (Fiore & Donath, 2004; Wen, 2015). Therefore, mobile dating apps have to incorporate various features to attract new users and keep their existing users (Fiore & Donath, 2004). This explains why all major Chinese dating apps have a live streaming feature that turns these apps into an entertainment portal.

On the specific ground, China has its own internet ecology. Part of the reason for this is that the Communist Party of China has banned Western internet services such as Google, Twitter, and Facebook from operating in the country. The ban has created a void for local services like Baidu, Weibo, and Renren (Fuchs, 2016). So far China has not officially forbidden Western dating apps such as OkCupid or Grindr in the country. However, because these Western apps are not offered in Chinese, they are not appealing to the locals. This has created a market niche for local dating app services. Chinese dating app companies learn from existing dating and social networking apps in other countries and create designs that are suitable to their Chinese market. For example, Tantan is based on Tinder's swiping style; Blued functions similarly to Grindr and Jack'd. Further, dating apps in China are heavily regulated by the Chinese government.

In March 2015, the National Office Against Pornographic and Illegal Publications fined Momo CNY60,000 (~USD8,600) and ordered it to remove group chats with explicit sexual discussions (see T. Liu, 2016). Zank, once a popular dating app for gay men, was shut down by the Office of the Central Cyberspace Affairs Commission in April 2017 for its live streaming of pornographic content. Rela was also temporarily taken off the shelves in 2017, allegedly because of its involvement in an organized protest for marriage equality in Shanghai. In April 2019, the government also took down Tantan for a couple of months because of its pornographic content. These complex relationships between Chinese dating apps, their Western counterparts, and the state have created a unique background for Chinese dating app culture.

Although the depiction above is derived from the Chinese context, these transformations are by no means unique to China. Similar transformations of intimacy and of dating technologies have occurred in other countries and cultures.[16] In the concluding chapter of the book, I revisit the notion of networked sexual publics and illustrate how this concept is useful in understanding the global emergence of dating app cultures.

RESEARCHING THE EXPERIENCES OF DATING APP USERS IN GUANGZHOU, SOUTHERN CHINA

Most of the evidence in this book comes from in-depth interviews I conducted in 2016 and 2018 with sixty-nine dating app users during two separate field trips to Guangzhou, a major city in southern China. Below, I briefly describe the location of my field trips, my informants, my interview protocol, and my analytical procedure. The appendix elaborates on the recruitment and interview processes and presents detailed information on each informant.

Recent research in China related to gender and sexuality has mostly been conducted in Beijing or Shanghai (e.g., Farrer, 2002; Kam, 2013; Pei, 2013; Wallis, 2015). Only a couple of studies have included Guangzhou in their multisited fieldwork (see Bao, 2018; Kong, 2011). Because I am reaching beyond Beijing and Shanghai, cities which non-Chinese readers may be more familiar with, some background information about Guangzhou may help put my informants' experiences in context. As the capital of Guangdong province, Guangzhou had around fifteen million permanent residents in 2018, among which approximately 62 percent were *hukou* 户口 (household register) holders and 38 percent were migrants. Guangzhou is the third largest city in China in terms of population after Shanghai and Beijing (Guangdong Bureau of Statistics, 2019). Over 98 percent of the city's Chinese residents identified as Han Chinese (Xing, 2011). Economically, the city is building strategic hubs for international shipping, aviation, and technology. Its regional gross domestic product reached CNY1,961 billion (~USD283 million) in 2016, an increase of 8.2 percent from the previous year, which was higher than the national average of 6.7 percent ("Guangzhou changzhu renkou 广州常驻人口," 2017). It is not difficult to find global luxury brands such as Louis Vuitton, Chanel, and Hermès and Western fine-dining restaurants in the city, particularly in Tianhe district. Culturally, Guangzhou is the origin of the Lingnan culture with its unique dialectic (Cantonese),[17] art (such as Cantonese opera), and food culture (such as dim sum).

The city is close to Shenzhen, one of the first special economic zones to open to marketization and foreign investment during the 1980s. It is also close to Hong Kong, where I was born and raised. Hong Kong is a former British colony, from which many Western ideas have been introduced into China. Historically, since the Qing dynasty several important social reformers have come from Guangdong province, including Zheng Guanying, Kang Youwei, and Sun Yat-sen (Song, 2016). Therefore, it is not an exaggeration to say that Guangzhou has always been open to new ideas and cultures. Internet penetration rate of Guangdong province was 74.0 percent in 2016, much higher than the national average of 53.2 percent (China Internet Network Information Center, 2017). According to Guangdong Communication Administration (2019), as at November 2019, for every 100 people in Guangzhou, there were 147.0 mobile phones. Specifically, in the city of Guangzhou, there were 32.3 million mobile phone users, among whom 29.3 million were using 3G/4G connection.

Guangzhou is a relatively international city for China. It has retained its historical, Lingnan way of living and is actively incorporating Western consumerism. In the neighborhood where I stayed during my second field trip, there was a take-away dim sum store run by an old Chinese woman next door to a juice store whose owners were two young Caucasian men from California. I could easily get a CNY20 (~USD3) lunch set with a bowl of congee and a rice noodle roll and a CNY60 (~USD9) gelato. People mainly speak Cantonese and Putonghua, but in Tianhe district, English can be used in commercial entities.

Of the sixty-nine informants, nineteen identified themselves as straight women, fifteen as queer women, sixteen as straight men, and nineteen as queer men.[18] Table 1.1 summarizes the demographic information of my informants at the time of their interview.

The criteria for becoming an interview informant in this study were as follows: being eighteen years old or older, living in Guangzhou, having never worked at companies that run dating apps or dating websites, and logging onto dating apps at least several times a month.[19] Because I wanted to recruit users from different backgrounds, I implemented multiple methods and modified my methods along the way. I recruited my informants through dating apps (Aloha, Blued, Momo, and Tantan) by using a "researcher's profile" that clearly identified me as a researcher and explicitly stated the purpose of my research. I also attended a public lecture on women's sexuality, and with assistance from the lecture's organizer and the speaker, I handed out leaflets to the audience. In addition,

Table 1.1 A summary of the informants' demographics ($n = 69$)

Sexual orientation and gender	Age range	Relationship status
Straight women ($n = 19$)	21–38[a]	Single: 12 Dating: 2 Married: 3 Other: 2
Straight men ($n = 16$)	19–37	Single: 8 Dating: 5 Married: 3
Queer men ($n = 19$)	19–28	Single: 14 Dating: 3 Other: 2
Queer women ($n = 15$)	18–34	Single: 6 Dating: 9

Note: a. One informant declined to disclose her age. She appeared to be in her forties.

I contacted two LGBTQ organizations in Guangzhou, which helped me reach out to their members and volunteers. Finally, some informants invited their friends to participate in the study.

The interviews took between forty-five minutes to two hours and were conducted in Putonghua or Cantonese. Of the sixty-nine informants, all but one consented to having the interviews be audiotaped. For the one informant who declined to be audiotaped, I wrote extensive notes during the interview. The interviews were semistructured. The core questions remained the same across all interviews. I began by asking the informants easy questions: "Which dating apps are you using now? Which one is your favorite?" Most people did not have difficulty expressing what they liked and disliked. This question helped the informants think about their experiences; it also helped me understand the one or two apps that my informants appeared to be the most invested in. Sometimes, the informants recruited from one app said their favorite app was another. In this sense, my approach is akin to the "open touring invitation" approach in the "media go-along" methodology proposed by Kristian Jørgensen (2016). In this "tour," I invited my informants "to narrate a pathway through the app that is mostly of their choosing" (p. 40). I asked questions such as the impressions they had of the dating apps before they started using them and the reasons they had downloaded these apps. Gradually, I asked questions about their use of the apps, including how they presented themselves on the profile and how they interacted with others. I always included one reflexive question at the end: "What do you think you would miss if your favorite app was gone tomorrow?" Although I had prepared a list of questions on key issues, I also allowed the informants to share the experiences they found meaningful.

Interviewing is a social process. My intersectional identity—being a cisgender gay man from Hong Kong and having been educated in the United States—may have helped me gain access to stories from these informants that they might not have otherwise shared, but it might also have made them reluctant to share other stories. I am aware that every story is told, framed, and interpreted from a particular position (Haraway, 1988). Therefore, the purpose of conducting interviews was to understand the significance, or the social meanings, dating apps had in the lives of the informants rather than to seek objective facts. As the informants' recounting of their experiences was necessarily influenced by my presence and probing, I, as a researcher, also participated in the coconstruction of their narratives. Further, my research was informed by feminist inquiry—in particular, standpoint epistemology (Harding, 1993). Standpoint epistemology posits that women in certain marginalized social positions, or standpoints, are able to generate more accurate knowledge. In my research, I privileged the narratives of women and queer people because they have experienced social oppression from patriarchy and heteronormativity in China.

All recorded interviews were transcribed. With NVivo 11, I used a two-cycle coding process (Miles, Huberman, & Saldaña, 2014). I created the first cycle of codes by describing each paragraph, paying attention to any references that might have been related to interpersonal, social, and political dynamics. Examples of these codes included "looking at handsome men" (from a heterosexual woman), "showing wealth in profiles" (from a heterosexual man), and "no lesbian characters in mainstream television" (from a lesbian woman). The second cycle of coding began with organizing the first cycle of codes. Codes from each transcript were constantly compared until some themes gradually emerged. For example, interview excerpts coded as "looking at handsome men" and "pondering the relationship between love and sex" were grouped together under the theme of "a laboratory of sexual experiments."

ORGANIZATION OF THIS BOOK

Larissa Hjorth (2008) demonstrates that, in order not to resort to a pedestrian cross-group comparison, we must attend to specificities of each group of users. In her case, they were female mobile phone users in Japan, South Korea, Hong Kong, and Australia, respectively. This book explores networked sexual publics of both genders and different sexual orientations. In considering what the best way to analyze and present my data, I followed Hjorth's approach. After this introduction, I present four chapters, each dedicated to one of the four networked sexual publics—straight women, straight men, queer men, and queer women. One may criticize such an organization for risking the compartmentalization of sexual experiences and reinforcing the socially constructed dichotomy of men in relation to women and straight in relation to queer. I would argue that the dedicated discussion of each of these four groups of dating app users allows me to dive into the intersectional experiences of each one to discover the unique struggles, dilemmas, opportunities, and challenges each group faces and to draw connections between these different groups when appropriate. Likewise, in presenting my data, I could have written each chapter in the same structure, covering the same topics, and using the same subheadings. However, doing so would not only inevitably produce repetitive content but also assume that issues pertinent to one group are equally relevant to another. Instead, my organization of each chapter reflects what was important to my informants, as reflected by the data.

Earlier, when I elaborated how technology is related to gender, I mentioned technofeminism. Developed within the parameters of science and technology studies, technofeminism suggests that technology is "both a source and a consequence of gender relations" (Wajcman, 2006, p. 15). To fully understand the relationship of women with technology requires an examination of women's interactions with information and communication technologies in daily life. In chapter 2, I ask the following: what do dating apps mean to straight female dating app users, and what challenges do they face when using these apps, in view of the status of women in contemporary China? In this chapter, I analyze how women interpret their use of dating apps. I argue that although dating apps may be a feminist tool, they conceal the structural gender inequality embedded in society at large.

Hegemonic masculinity refers to idealized forms of masculine behaviors relative to the behaviors of women and other subordinated masculinities such as gay, working-class, and racial minority masculinities (Connell, 1987). However, masculinities are not inherent qualities of men; they must be learned and performed. Accordingly, in chapter 3, I take a constructivist approach to examine the ways in which Chinese straight men perform gender on dating apps (Butler, 1999). Specifically, I look at their interpretations of the apps, self-presentations on the apps, and interactions with women they meet on these apps. I argue that their performances are best understood relative to *wen-wu* 文武 (literary-military) masculinities, an indigenous Chinese concept developed by Kam Louie

(2002). My analysis shows that although some gender performances of my male informants appear to be inclusive of femininity, they also reproduce the existing gender inequality.

Chapters 4 and 5 bring readers to the lived experiences of queer communities in urban China. When I spoke with my queer male informants, I quickly noticed a phenomenon that has been less prevalent among other groups of informants—constantly deleting and installing the same app. Instead of taking a cognitive approach that assesses the usefulness of a technology (F. Davis, 1989; Brubaker, Ananny, & Crawford, 2016), in chapter 4, I turn to the affective aspect of dating app use. Emotions move people. They pull people closer to some things and push them away from other things (Ahmed, 2004b, 2010). I argue that the cycle of deleting and installing among queer men is a manifestation of the contradictory affects my informants had in relation to their experiences with dating apps. I differentiate two types of emotions. In-app emotions are directly derived from the everyday use of dating apps. Out-of-app emotions are rooted primarily in the way male homosexuality is treated in contemporary Chinese society. I show how the everyday use of dating apps and contemporary queer politics have generated both positive and negative emotions related to dating app use.

I devote chapter 5 to the experiences of queer women in China. I view queer women as sitting at the intersection of two oppressed identities (Crenshaw, 1989). Because they are both queer and women, they face greater challenges than straight women and queer men combined. In this chapter, I ask the following: how do popular lesbian dating apps contribute to the queer world-making project and connect queer women to their community? I look at the features of two popular dating apps and the users' perceptions of them. I then propose the affordance of communal connectivity to account for the missing communal aspect in the typology of affordances I discussed earlier. At the same time, I recognize that despite their potential to connect queer women with their community, these apps reinforce heteronormativity within the community. Thus, they were—as we all are—prevented from reaching the queer utopia (Muñoz, 2009).

Chapters 2, 3, 4, and 5 are each adequately contextualized in a historical background on the relevant debates and issues related to gender and sexuality in China. Readers are, therefore, welcome to read individual chapters if they are interested in a single gender or sexual category.

In the concluding chapter, I synthesize the findings of the previous chapters and provide a more thorough theorization of networked sexual publics, putting this concept in conversation with the three theoretical anchors I discussed above. Based on the research presented in this book and my reflections on some of the latest developments on dating apps worldwide, I put forth several propositions concerning networked sexual publics and suggest ways for scholars to further investigate this emerging global phenomenon. With this concept, I hope societal discussions of dating apps will move beyond the topic of hookups.

2

ARE DATING APPS A FEMINIST TOOL? A TECHNOFEMINIST ANALYSIS

Every weekend, hundreds of parents conglomerate in one section of Tianhe Park to help their adult child, usually a single child due to the one-child policy, look for a spouse. In this park, they literally "advertise" their children to other parents. Parents fill out an A4-size form entitled "Weihun Qingnian Jiazhang Xiangqin Huodong Guanggao Biao 未婚青年家长相亲活动广告表" ("Unmarried Youth's Parents Matching Activity Advertising Form").[1] This form documents their child's age, height, weight, education level, property ownership, household register, income, nature of occupation, and spousal selection criteria. Hundreds of forms, hung on strings and tied around trees, are grouped in sections according to the gender and age of the child. People with an overseas degree have a separate section to designate their increased status. Worried parents walk around and look at the other forms or stand beside their own forms, waiting for other parents. In their analysis of a similar activity in People's Park in Shanghai, Jun Zhang and Peidong Sun (2014) argue this kind of parental intervention does not revive the traditional arranged marriage practice. Instead, they see it as a response to the contemporary discourse that people, particularly women, have a hard time getting married if they become too old. But how old is too old?

Nancy, introduced in the last chapter, was showing me Tianhe Park as we chatted. But she became quiet when we started reading the forms.

Suddenly, she turned to me and asked, "Do you think *shengnü* like me will have a better market in the United States?" The term *shengnü* 剩女 (leftover women) emerged in the media discourse of the mid-2000s to refer to older, educated, rich, single women (Hong Fincher, 2014; To, 2015). It is a pejorative term. Women do not wish to be addressed as *shengnü*, and parents do not want to hear their daughter being described as such. Underscoring its sexist subtext, no equivalent term exists for older, rich, single men. Instead, these men in China are known as *zuanshi wanglaowu* 钻石王老五 (diamond bachelor)—a term that does not carry a negative connotation.

Matching activities like this reflect gender inequality in Chinese society that places women in an inferior, subordinate position. Besides the different connotations for single women and men, gender inequality also manifests in dating and romantic practices. The traditional norm in American dating is that men are expected to lead the courtship process. Women who make the first move are considered "too easy" and are not valued (Bailey, 1988). The same convention to give more control to men exists in China, where women are taught to be reserved when seeking romance. In response to this gendered phenomenon, some Western dating apps have strived to give more power to women. Bumble, founded by Whitney Wolfe—a cofounder of Tinder who later quit the company due to the sexual harassment she faced there—was designed in a way that a conversation between a woman and a man could be initiated only by the woman. In doing so, Bumble has endeavored to, as Wolfe puts in, "reconfigure the way that we treat each other" (cited in Tait, 2017, para. 10). Once, an app established in Europe, similarly lets women rate their male dates ("Dating app Once," 2018), further disrupting established gender norms.

In this chapter, I explore the following question: do dating apps empower women in China? My concern with empowerment stems from Linda Layne's (2010) suggestion that the ultimate feminist concern is reclaiming women's autonomy. Accordingly, technologies can be feminist if they are consciously designed to make women's lives easier or their unintended consequences improve women's lives. Deborah Johnson (2010) specifies four forms that feminist technology can take—technology that improves the condition of women, technology that contributes to gender equality, technology that favors women, and technology that elicits more equitable gender relations than those associated with prior technology. However, whether a technology can become feminist cannot be disconnected from women's status in society.

In order to answer my own question, in what follows, I first sketch the trajectory of women's status in modern China. This sketch is partial, but I hope it captures the specific segments of the trajectory that provide sufficient contextualization for the later discussion on the role of dating apps. As I mention in the introductory chapter, my analysis of the use of dating apps is informed by technofeminism's conception of gender and technology as mutually constitutive. This approach pays attention to the gendered meanings related to the use of technology (Wajcman, 1991, 2006, 2007).[2] The remainder of this chapter tells the stories of my straight female informants. Through these stories, I analyze the possibilities and challenges dating apps have presented and discuss why dating apps ultimately may not be able to fulfill the political goals feminists strive for.

THE CHANGING STATUS OF WOMEN IN MODERN CHINA

Traditional Confucian culture considered women to be men's property. The "three obediences" (sancong 三从) in Confucianism demanded a woman obey her father as a daughter, her husband as a wife, and her sons after her husband died. Marriage was arranged by parents as early as six or seven years old. For a daughter to be married, her parents had to pay a dowry to the prospective husband's family. Thus, for parents, having a daughter was a net loss for their family's wealth. Furthermore, once married, there was no way for a woman to request a divorce (Fei, 1939). Similar to traditional Western societies, Chinese women were not given any sexual agency. If they exercised their sexual desires, they were stigmatized as a dangfu 荡妇 (loose woman), while no equivalent terms were applied to men who had extramarital encounters. Women in China during these years followed scripted gender roles or risked the consequences.

In 1950, one year after the Communist Party of China (CPC) founded modern China, the party banned arranged marriages and allowed "no-fault" divorce if mediation by the state failed (D. Davis, 2014). In 1954, a consistent policy regarding women's voluntary use of contraceptives was also formulated. After this point, contraceptives became available without a doctor's prescription (White, 1994). Meanwhile, the parent-child relationship also underwent a radical transformation. Instead of the father being the authority in the household, the children occupied the same position as their father, unless the father was a cadre in the CPC (Yan, 2009). This shift in policy cultivated autonomy and self-development among young women. The image of "iron girls" (*tie guniang* 铁姑娘) was particularly propagated during the Cultural Revolution. The iron-girls campaign was used to mobilize and organize women to enter traditionally male occupations. In 1964, Mao Zedong stated that "times have changed, men and women are the same. Whatever men comrades can accomplish, women comrades can achieve as well" (cited in Jin, Manning, & Chu, 2006, p. 617). During this period's legal and policy reforms, women experienced an increase in social status that had been unheard of in premodern China. Zheng Wang (2005) calls this series of efforts by the state to improve women's status "socialist state feminism."

However, one should not assume that, with state feminism, women in China during the Maoist era had achieved gender equality. The All-China Women's Federation, founded in 1949, is the designated leader of women's movement in the country. It is not a grassroot organization but is part of the CPC. While the federation's members are placed into the state structure at every level (Judd, 2002), the fact that the federation is part of the CPC but not of the government renders the federation power-less because its members are not allowed to play a leading role in governance (Z. Wang, 2005). Further, during the Maoist era, by absorbing the women's movement into its anticapitalist agenda, the CPC suppressed sexual differences (Rofel, 2007; J. Yang, 2011). It was not until the economic reform that women's bodies were allowed to be sexualized. In the late 1970s, images of fashionably dressed and sexually appealing women began to appear in the state-run media (H. Evans, 2008).

In recent years, some scholars have observed that urban women in China continued to develop their sexual agency and marital power. Yuxin Pei (2013) documents how, beginning in the early 2000s, several female writers gained fame by writing erotic literature. This was the first time since the economic reform that the sexual imaginations of female writers dominated the public sphere. Through extensive interviews with straight women living in Shanghai, Pei also found they had diverse sexual and romantic experiences. Some of these women stated that with their sexual capital, they had greater privileges than men. Pei argues that the women she interviewed "used 'gender equality' as a weapon, as a tool, not to strive for 'equality between men and women' but to take advantage" (p. 189, my translation from Chinese). During this same time, Susanne Yuk-Ping Choi and Yinni Peng (2016) also found a gradual change in the gender hierarchy among rural-to-urban migrant couples and families. The wives in these households were expected to control their husband's finances and had greater bargaining power in deciding where the couple would eventually live.[3]

Nevertheless, existing side by side with these progressive trends, the CPC has reregulated women's lives. Among these regulations, the one-child policy, implemented from 1979 to 2015, created a severe national sex ratio imbalance. Chinese families traditionally prefer sons over daughters because sons inherit the family's name and are expected to make money for the family. Under the one-child policy, most families were fined for having additional children. The preference for sons together with the threat of financial penalty for having two or more children have led to many cases of female infanticide, selective abortion, and unreported birth (Hull, 1990; Johansson & Nygren, 1991; Junhong, 2001). According to China's 1990 population census, for every 100 newly born baby girls, there were 119.92 baby boys. The 2010 population census revealed a further deterioration of the imbalance: the ratio increased to 100 to 121.21. In Guangdong province, the ratio reached 100 to 129.29.[4]

The government has treated this imbalance as the cause of the rise in sex-related crimes because many marriage-age men cannot find a wife (Li, 2014). Men have fewer choices because women enjoy more education and vocational opportunities than before, while men prefer wives with less education and fewer accomplishments than themselves. An increasing number of women with advanced education and established careers have remained single. The term *shengnü* gradually emerged in media discourse starting in the mid-2000s (Hong Fincher, 2014). The exact origin of the term is difficult to trace. However, in 2007, this colloquial term became one of the 171 official new words highlighted in *Zhongguo Yuyan Shenghuo Zhuangkuang Baogao* 中国语言生活状况报告 (*The Chinese Language*

Life Report) published by the Ministry of Education and the National Language Committee (Zhang & Sun, 2014). According to this report, the official definition of the term is "Urban professional women who are over 27 years old who have high education level, high salary, high intelligence, and attractive appearance, but also very high expectations of marriage partners, and hence are 'left behind' in the marriage market" (cited in To, 2015, p. 1). Popular television shows in China have also blamed women for being too ambitious and career-minded (Feldshuh, 2018). However, as the national statistics have shown, because there are more men than women, it is *men* who are "leftover," not women. Therefore, Leta Hong Fincher (2014) contends that the term was created by the state to coerce women into marriage:

> In one sense, "leftover" women do not exist. They are a category of women concocted by the government to achieve its demographic goals of promoting marriage, planning population, and maintaining social stability. The state media campaign against "leftover" women is just one of the signs that in recent years, contrary to many claims made by mainstream news organizations, women in China have experienced a dramatic rollback of rights and gains relative to men. (p. 6)

This brief social history illustrates the dynamics between gender structure, gender symbols, and gender identity related to women (Harding, 1986). Society has organizing principles that structure relationships between men and women through kinship and economy. Gender, reflected in images of manhood and womanhood, is symbolic. On the personal level, gender is also an identity. It is about who one is and guides how one should present oneself to others. These three aspects of gender may reinforce or contradict each other. In the Chinese context, structurally speaking, women have risen from the submissive position dictated by the "three obediences" in ancient China to a position closer to their male counterparts during the heyday of socialism.[5] A distinct feminine identity has also been cultivated by consumerism since the economic reform. Nevertheless, we have witnessed the emergence of *shengnü* as a pejorative symbol for educated but unmarried women in recent decades.

Under this context, what is the role of dating apps in shaping women's position in Chinese society? Can these apps improve the condition of women and contribute to greater gender equality? And are there any unintended consequences of using dating apps that may elicit more equitable gender relations than those associated with prior infrastructure of intimacy, like *xiangqin* 相亲 (matching) in the park? In the following, I present four metaphors of dating apps drawn from interviews with straight female informants. I also report challenges my informants faced in their experiences of using dating apps and the ways they reacted to these challenges.

SIGNIFICANCE OF USING DATING APPS

Extending the social shaping of technology paradigm (MacKenzie & Wajcman, 1985), technofeminism focuses on women's technological practices and the implications these practices have on their status in the society (Wajcman, 1991, 2006, 2007). Technofeminism is also concerned with women's agency and their technological interpretations. Several metaphors emerged from the narratives my informants provided that account for their experiences using dating apps such as Momo and Tantan. While I present these metaphors separately, they are not mutually exclusive. Some of my informants held two or three interpretations at the same time. I intend my typology presented here to showcase a wide range of interpretations across four central metaphors—a laboratory for sexual experiments, a springboard to romance and marriage, a third place between home and the workplace, and a gateway to new worlds.

A LABORATORY FOR SEXUAL EXPERIMENTS

David Gudelunas (2008) argues that for many years in the United States, newspaper advice columns provided knowledge on sexuality to the public during a time when the topic was forbidden in schools. Some of my informants pointed out that the education system in China has similarly provided little room for its citizens to think about intimate relationships or sexuality. Thus, dating apps offer an interactive environment in which women can explore their sexuality and think about the relationship between sex and love. Kangqi, age thirty-four, came to Guangzhou from a small city in Sichuan province three months before our interview. She complained, "Our education . . . does not teach you [men] how to handle intimate relationships, how to get along with women, how to express affection, what is or what isn't sexual harassment." She had been looking for open relationships in recent years, and Momo had played an important role in her experiment:

> [In my town,] there were only some [Momo] users. . . . I and one of the men developed a casual sexual relationship, a purely sexual, mutually gratifying relationship. Although I was living in a small town, I was relatively open-minded.

She said that because none of her boyfriends lived in Guangzhou at the moment, she planned to use Momo to look for boyfriends.

Other informants' stories showed that their experiences with dating apps provided them with opportunities to reflect on their attitudes toward sex and relationships. Xiaojiao, age thirty, recalled meeting a man on Tantan who was in an open relationship with his girlfriend. She said, "In the past, I knew nothing about sex. He was my teacher. I think that I learned from him how to separate sex [from romantic relationships]." Like Xiaojiao, Queenie, age twenty-five, met her sex partners on Tantan. But when I asked her if she could disentangle sex from emotion, she said:

> No. My point is, I enjoy the process. But first, I have to think [the person] is okay, is interesting. [A stranger] can bring me a sort of fantasy, like the feeling of smoking weed. . . . But if you are talking about sex without love, it doesn't work. I have to strike a balance.

Thus, although Xiaojiao recognized she could enjoy sex without committing to a person, Queenie understood that what gratified her was the liminal space between familiarity and strangeness. I refer to dating apps as a "laboratory" because for Xiojiao and Queenie, they provided a space to test out how their sexual desires could be acted on.

Pushing these experimentations even further, younger informants tended to use dating apps for fantasy. As a Tantan user, Brady, age twenty-three, said, "If I feel bored, I will log onto the app to see if there are any good-looking people nearby, check out their profiles, and 'like' their profiles if they are good-looking." Amanda, age twenty-four, said she did not believe that dating apps were places to look for relationships. Eventually, she became a pure lurker who just looked at pictures. Rosy, age twenty-one, also enjoyed just looking at the photographs. She did not bother wasting time chatting with men: "I just press the 'like' button. That's it." These practices reflect Brady, Amanda, and Rosy's interests in experimentation with browsing media that may not lead to dates or long-term relationships.

Moreover, Rosy's refusal to waste time also reflected a cost-and-benefit analytical mentality. Dating apps serve as a suitable laboratory for sexual experiments not only because they provide experimental "materials"—a diverse group of men—but also because they demand very low involvement. Rosy appreciated the efficiency of using dating apps to look for hookups because they merely required a photograph to be uploaded and a couple of lines to be written on the profile. To her, dating apps required little emotional investment: "If they work, I will keep using them; otherwise, I will let go. It is like a game. I won't be very serious." As Zygmunt Bauman (2003) wittily comments, "termination on demand—instantaneous without mess, no counting losses or regrets—is the major advantage of Internet dating" (p. 65). I see young women like Rosy fully embracing the advantage of easy termination in relationships on dating apps and the use of these apps itself.

A SPRINGBOARD TO ROMANCE AND MARRIAGE

Going against the discourse of dating apps as *yuepao shenqi* 约炮神器 (a magical tool for hookups), some informants said they believed they could find a boyfriend or husband on these apps. Katie, age thirty-three, was one of the few informants I interviewed who had used Western apps such as OkCupid because of her good command of English. She used OkCupid to look for long-term relationships, even though she said many people on the app were seeking hookups. Dating apps provided Katie with a new set of potential boyfriends because she described her social circle as limited: "In gatherings, my friends will show up with some new friends, but these people are usually married, so I cannot look for a partner on such occasions."

Other Chinese urbanites were also interested in expanding their pool of potential boyfriends. Jessica's strategy was to get to know as many interesting people as possible on Momo and then, ideally, to develop a serious relationship with one of them. She emphasized, "I have a clear goal; after all, I am at my current age." As I discussed earlier, she was responding to the enormous social pressure to get married to avoid the psychological threat of being called *shengnü*. As a twenty-seven-year-old college graduate employed as a civil servant, Jessica just made the cut of becoming a *shengnü* according to the official definition. She complained that whenever she called home, the first question her mother asked was whether she had a boyfriend. It is no wonder that she was concerned.

Some features afford a safer way to seek a suitable partner. Queenie, who looked for sex partners on Tantan, enjoyed reading the personal profiles created by men that allowed her to better understand them. She also set up an age filter so that only people between eighteen and thirty years old showed up on her app. Nancy, age twenty-eight, appreciated the alert function on these apps, which protected her from sexual harassment. "Because [the app] has a list of preselected keywords—for example, *'yuepao'*—the system will remind you to report [harassment]." Polly, age twenty-nine, showed me how this automated alert system worked on Momo. She sent me a message containing the word *yue* 约. On my end, I received a message from Momo reminding me to report if I found the message from Polly offensive (figure 2.1).[6] Overall, my female informants interpreted these features—including detailed profiles, filters, and the alert system—as beneficial. These features were helpful to their search for Mr. Right, who could save them from being stigmatized as *shengnü*.

A THIRD PLACE BETWEEN HOME AND THE WORKPLACE

To some of my informants, dating apps are what Ramon Oldenburg and Dennis Brissett (1982) call "third places." These are places that "exist outside the home and beyond the 'work lots' of modern economic production . . . where people gather primarily to enjoy each other's company" (p. 269). This interpretation is particularly salient for my nonlocal informants.

Xiaoshan, age twenty-nine, moved to Guangzhou two years ago so she could work to support her husband and son back home in Guangxi province. Working as a masseuse in a hotel, she said the nature of her job made the workplace very competitive. Therefore, she was not close with her colleagues. When she wanted to vent, she turned to Momo because

Figure 2.1
After my informant sent me a message containing the word *yue* 约 on Momo, the system autogenerated a message reminding me to report her if I found the message offensive. (Screenshot taken by the author on October 17, 2016)

only strangers could see her "status updates." She explained, "I cannot share how I feel on WeChat because my family and friends will be worried if they see these messages."[7] Using Momo and hanging out with people she met on the app were her sole social activities. She liked hiking, and she joined groups that were related to hiking and outdoor activities in the "groups nearby" feature in Momo. Personally having moved to a new city for study and work, I was glad to hear that she was able to find hiking buddies on the app. Xiaojiao, who had also left her hometown, told me that "My social life is my life on Tantan." Similarly, when Jessica first started working for her company in another city, most of her colleagues were married men who often stayed home on weekends. Momo was the primary channel for these women to make friends.

While home was simply unavailable to my nonlocal informants like Xiaoshan, it also could not provide the kind of sociability my local informants needed. Due to the former one-child policy, many younger Chinese have no siblings. Polly, a local Guangzhouese, told me she just needed someone to talk with. Classmates might not have been the best listeners because they were in a competitive environment. However, on Momo she befriended two men and one woman. She remarked:

> They are around my age. Our friendship was formed in a very short period of time during a gathering. There is chemistry, not the chemistry between a boyfriend and a girlfriend. It feels like we have known each other for years.

Further, some users are torn between their workplace and their household. For Jennifer, a thirty-seven-year-old single mother, her time was divided between her job and her nine-year-old son. On weekdays, she was either at work or with her son. She devoted the weekends to her son, unless her former husband took their son out. "The social circles of modern people are tiny," she lamented. She did not feel comfortable revealing to her colleagues that she was a single mother because being a single mother, like being a *shengnü*, was stigmatized. However, on Momo, she could present herself as a single mother. She met two other women who were also mothers from the chat group supported by the app. When she had spare time, she hung out with them. As a local Guangzhouese, she also set up a chat group on Momo for Cantonese speakers. I joined the group with her permission. Although the group's chat name contained the terms "movie" and "tabletop games," during the entire duration of my fieldwork in 2016, the group members never met face to face for movies or tabletop games. All interactions happened on the app. The conversations centered mainly on daily life. On one occasion, a man flirted with a woman, inviting her to his home. The woman questioned his "size." The conversation ended with both of them admitting they were simply kidding. Jennifer told me that this kind of sexual joke was quite common in her chat group but no one took this type of conversations seriously. Nonetheless, she saw these trivial jokes as a welcome distraction from her life, which was mostly devoted to work and her son.

The above narratives show how dating apps have created a third place for my informants where they could enjoy another's company. Although this space was sometimes filled with sexual jokes, it was highly social in nature. The emotional bonding enabled friendships to form. This image of a third place is not commonly found in the narratives about dating app cultures in the West.

A GATEWAY TO NEW WORLDS

Like social media, dating apps attract people from all walks of life. My informants had more opportunities to interact with people and try out alternative experiences by using dating apps. Wenwei, who declined to reveal her age, started using Tantan at the beginning of 2016 after her divorce. She said that her life "couldn't have been livelier since then." Katie provided me with more details of her experience:

> In the past, I did not go to bars. But after chatting with [people on OkCupid], I have started going to bars with friends after work. I want to try new stuff, different things. There are so many different types of people with different lifestyles, I want to see more.

Meeting more people made Katie think about studying overseas to improve herself. An even more inspiring story was that of Xiaolan, who was twenty-three. Among all the straight female informants I interviewed, dating apps had the most profound impact on her personal life:

> I once met a man from Slovakia. He said to me, "You don't know anything." I then read the BBC news like crazy, to gain a wealth of information. I have never been so curious about the world, never been so enthusiastic about knowledge.

These examples illustrate how female dating app users accumulate social capital from dating app use. Social capital is the actual or virtual resources a person can accrue from social relationships (Bourdieu & Wacquant, 1992). There are two kinds of social capital (Putnam, 2000). Bridging social capital is formed by wide but relatively shallow connections with people from different backgrounds who can provide diverse information. Bonding social capital is based on strong relationships that offer trust and emotional support. Prior studies have shown that the use of social media contributes to both bonding and bridging social capital (e.g., Chen, 2011; Zhang, Tang, & Leung, 2011). The narratives of my informants—for example, "My social life is my life on Tantan" from Xiaojiao and "I want to try new stuff, different things" from Katie—demonstrate that dating apps are a source of both bonding and bridging social capital.

CHALLENGES IN USING DATING APPS

Technology almost never has a one-sided influence on us. Julie Frizzo-Barker and Peter Chow-White (2012) found that working mothers benefited from the flexibility, efficiency, and connectivity afforded by smartphones. Simultaneously, they had to handle the stress that came with the "always on" lifestyle, complicating the way they reconciled their private identity as mothers and their public identity as workers. Although dating apps open up new possibilities for women, they also introduce new challenges. Interviewing female app users revealed three major challenges they faced and the measures they took to tackle these challenges.

RESISTING THE STIGMA ASSOCIATED WITH DATING APP USE

Research has shown that "slut shaming" is common in American dating app culture (Birnholtz, Fitzpatrick, Handel, & Brubaker, 2014; Lee, 2019). What is different in China is that simply using a dating app can jeopardize one's reputation. Katie never mentioned to her friends that she was an OkCupid user: "In China, people like judging others. . . . If I use this app while others don't, they will question my intention. 'Why do you use this app to meet people?' They will think you are weird." The labeling of *yuepao shenqi* also contributes to this stigma. Like Katie, Jessica never said a word about using Momo to her friends. "I feel that if I mention Momo, people will basically associate it with *yuepao shenqi*. I will feel embarrassed, even though my major objective is to make friends." The feeling of embarrassment has to be understood against the cultural assumption that women should be passive and subservient. Deviation from this norm brands a woman a *dangfu*. Further, although Jessica did not want to be judged by her colleagues, when she told me she had come across a male colleague on Momo, she said it with a mix of teasing and disdain. This reaction perhaps represents the internalized stigma associated with using dating apps.

Some informants experienced firsthand negative judgments from others. One time, Nikki, age twenty-six, went on a date. The young man noticed Momo on her phone and questioned why she was using, in his words, "a thing like that." Nikki explained to him that she downloaded the app simply because her friends were using it. She also had to delete the app in front of him to prove her innocence. Another case was Xiaolan. Her Tantan profile said, in English, "We can talk everything, including life, love, or something other [sic]." She told me she used to have a few sentences describing her sexual openness, but she had to delete them because some men on the app criticized her promiscuousness. Not every woman I interviewed hid her dating app use. Nonetheless, they selectively disclosed the information they believed their family or friends would accept. For instance, Xiaoshan told her husband back home that she used Momo, but she did not tell him that she hung out with people she met on the app. Xiaolan also told her roommates about Tantan, but she did not reveal the sexual encounters she had had.

These small denials hint at a deeper undercurrent in Chinese society. Although women are gaining sexual independence on a macro, societal level, as illustrated by the sexualization of female bodies and the rise of several female erotic writers I discuss earlier in this chapter, the scenarios above show that women's sexuality is still under micro, interpersonal patriarchal control. Expressing and exercising sexual desires are still mainly reserved for men.

ASSESSING MEN'S PURPOSES

Affordances provide possibilities. Through various affordances of dating apps, users may pursue different types of relationships. Being proximate to each other enables an easy coffee date, but some users may exploit this affordance for casual sex. Thus, my female informants wanted to identify the purposes of the male users they encountered. To do this, they examined the men's photographs and written profiles and chatted with them.

Photographs are the dominant aspect of a dating app profile, which I call visibility (L. S. Chan, 2017b). My informants held different folk theories linking people's photographs and their relational goals. Wenwei told me she usually ignored profiles that contained only scenic photographs rather than faces. To her, the men behind these profiles were unwilling to disclose their identity and were not looking for a serious relationship. Nancy recounted an experience with a good-looking young man. She was interested in him until one night, she found that he had posted a shirtless photo. "Just looking at the photo made his objective very clear," she told me. "It was midnight. Obviously, he was looking for girls [to hook up with]."

My informants also found the length of the men's written profiles to be good indicators of their objectives. Katie said long and well-crafted profiles meant that the men were not into casual relationships. Conversely, "If nothing is written, I will find him unreliable," she supplemented. Polly agreed with Katie: "For men who use Momo for *yuepao*, they will not share their information. . . . If they are married and they share their authentic information there, . . . their family members might get involved."

Very often, a man's purpose cannot be identified from his static profile; his purpose may be best discovered through chat. Kangqi showed me her exchanges with a man she had added to WeChat. After asking if Kangqi was working, a benign topic, the man suggested going to her home and boasted about his sexual potency. Sometimes, men were less explicit in their sexual intent. Brady commented, "They give you a hint. . . . They usually suggest meeting at 10 p.m. It is too late for dinner or movies. So I won't accept the invitation." However, reading between the lines requires experience and practice. When she first became involved in the dating app scene, Rosy did not know what a single-word message *yue*? 约? (meet?) meant. She innocently thought it was referred to meeting up for coffee or for a movie. She later learned from Baidu, the Chinese counterpart of Wikipedia, that this was a coded word for hookups.

HANDLING SEXUAL HARASSMENT

The third common challenge my informants faced was handling unsolicited sexual requests and sexual harassment. All of the women reported being asked for hookups, regardless of which apps they were on. This phenomenon reflected both the public impression that these apps were *yuepao shenqi* and the increasing sexual openness of contemporary China. Dissociative anonymity—the opportunity to separate their online actions from their real-world identity—might also have encouraged some users to become more direct in soliciting casual sex from others (Suler, 2004).[8] These sexual solicitations were extremely direct, involving phrases like *yue bu yue*? 约不约? (meet or not meet?) or simply *yue*? as noted above. Even Kangqi, who used apps to look for sex, said that some men "are way too single-minded, pushy about hooking up with you."

Although some of my informants used these apps for sexual experimentation, others regarded dating apps as a platform for romance, friendships, and exploring a new world. To this end, they developed various tactics, from being proactive to passive, to respond to undesirable sexual solicitations. The following are the four tactics I gathered from my informants' experiences.

The first, a proactive one, was to preemptively indicate one's rejection of casual sex on one's profile. On her Momo profile, Polly wrote, *"Ni ai wo, ta gun; ni ai ta, ni gun; yuepao zhe, gun* 你爱我, 她滚; 你爱她, 你滚; 约炮者, 滚" (If you love me, she gets out; if you love her, you get out; asking for hookups, get out). Amanda's statement on Tantan was concise: *"Bu yue zhi liao* 不约只聊" (Not for hookups, just for chat). Because Tantan shows users how many people have "liked" their profiles, Amanda found that after putting this statement on her profile, she received fewer "likes" per day. Coco, age thirty-four, wrote, *"Ni gan re wo shishi* 你敢惹我试试" (Don't you dare proposition me) on her Tantan profile.

Were these statements effective? Yes, but not always. Although Polly, Amanda, and Coco reported fewer sexual requests after putting these statements on their profiles, Jennifer told me that her warning message did not decrease the number of sexually harassing messages she received. The failure of the first tactic led to the second one, which was to report harassers to the app. This tactic is reactive because women report offenders after receiving a harassing message. Commenting on Momo, Jennifer said, "The administrators . . . will delete [harassers'] accounts immediately if they receive complaints." The administrators, according to Jennifer, reviewed the past ten message exchanges between her and the suspected harasser. It took only a couple of minutes for the harasser's account to be removed. In her workplace, however, Jennifer and her female colleagues rarely reported sexual harassment: "It is inappropriate to report . . . unless you are thinking about quitting." What we see here is that being a female dating app user gave her more power than being a female employee in the workplace.

The third tactic was blocking or ignoring the harassers. While reporting harassers to administrators could result in their accounts being permanently removed, blocking or ignoring harassers allowed them to stay. This was not a passive tactic but an empathetic one because women who adopted it tended to agree that hookups were "normal." Fanny, age thirty-one, said, "It is a separate issue that I don't [hook up], but the existence of this practice is perfectly normal." If any of the men she met on Tantan implied they had sexual intentions, she stopped replying to them. Similarly, Yiping, age thirty-eight, ignored or blocked harassers because she said any reply provided them with positive reinforcement. She remarked that the hookup culture was

> Inevitable. . . . These young people do not need to go through the sexual liberation period. They do not need to be "liberated." They were born in an environment where [hookups] are as normal as drinking water.

The final tactic was a passive one: to quit. In their study of gay men quitting Grindr, Jed Brubaker, Mike Ananny, and Kate Crawford (2016) found that people left Grindr partly because the app was too much about casual sex and was therefore dehumanizing. Similarly, several women told me that they had uninstalled dating apps because of the disgusting messages they received. Chloe, age twenty-three, was hoping she could meet friends on Momo, but she deleted the app immediately after a man asked her for a hookup. Jennifer and Kangqi also uninstalled Tantan because there were too many, as Jennifer put it, "people like that" on the app.

LIBERATING OR DISCIPLINING?

When studying the social significance of mobile phones to young Chinese migrant women working in Beijing, Cara Wallis (2013) coined the phrase "immobile mobility" to point to a contradictory observation. Although young women have surpassed physical and social boundaries through their mobile devices, they have continued to occupy a relatively low position on the socioeconomic ladder. Wallis's analysis suggests that, to evaluate the feminist potential of a technology, we cannot isolate the technology from the larger sociopolitical context in which it is used. Based on the narratives of my informants, what should we conclude about the feminist potential of dating apps? Do dating apps reproduce or disrupt the existing gender inequality in the broader context of women in China? I argue that, on the surface, dating apps and their various affordances provided my informants with opportunities to disrupt the oppressive patriarchy. In the following discussion, I highlight three such disruptions. Nevertheless, I also emphasize that each of these progressive developments masks an underlying structural gender inequality.

First, dating apps, such as Momo and Tantan, provide a medium for people such as Xiaojiao and Queenie to explore sexual desires and assert sexual agency. As Pei (2013) describes, some female writers have developed a distinct gender identity by writing erotic novels. In this same register, my informants' sexual experiments on dating apps helped them develop their female sexual identity. These apps became an arena where they could negotiate and participate in hookups, an activity that was and is still widely perceived to be exclusive to men. Kangqi was pursuing open relationships through the dating apps. The pursuit of multiple sexual relationships in China, as Pei argues, is a manifestation of women reworking the traditional sexual script that stresses sexual exclusivity. In this sense, dating apps have created more equitable gender relations in terms of sexual exploration, one that was not available in prior infrastructure of intimacy such as *xiangqin*.

Even though dating apps seem to be a liberating space for women to develop their gender identity, they experience condemnation for using them. Recall that Xiaolan had to rewrite her Tantan profile to hide her sexual intent after being criticized by men on the very same app. As a result of social stigma, women had to keep their use of dating apps private. Using dating apps, regardless of the purpose, could invite harsh judgments from others. To reword Gayle Rubin's (1984) summary of sexual negativity, "dating apps are presumed guilty until proven innocent."[9] Nikki, who had Momo installed in her phone, was pressured to uninstall it in front of her date. What is under male surveillance now is not just women's sexuality but their use of technology.

Second, younger women, such as Brady, told me they enjoyed looking at pictures of men on dating apps, subverting the subject-object gender structure. The notion of the "gaze," a manifestation of power and pleasure, has been widely discussed in psychoanalysis and film theory. Instead of the classic arrangement in which male audiences look at female characters (Mulvey, 1975), these young women derive gratification from looking at men's photographs on their phones. When swiping or clicking, they are also judging the appearances of these men in a privatized form of a beauty pageant. Although they are at the same time being looked at by men, dating apps break away from the traditional unidirectional gaze of men at women, giving them opportunities that were not available in earlier eras. A similar subversion is observed by Lisa Wade (2017), who found that, in American campus hookup culture, female students rejected sexism by objectifying men back.

This temporary visual gratification, however empowering in the short term, has not altered the derogatory cultural discourse of *shengnü*. Brady and Rosy were too young to face parental and societal pressures to get married. Their youth allowed them to merely look at photographs of men without worrying about being single. Other informants, such as Jessica and Nancy, who were just three or four years older than Brady and others, had already confronted pressure from their family and relatives. Jessica complained: "People will ask [my parents], 'Has your daughter gotten married yet? Does your daughter have problems? How come she doesn't have a boyfriend at her age?'" Earlier research has shown how single women have resisted this stigmatized label of *shengnü* (Gaetano, 2014). Some emphasized their career ambitions, while others said they believed in the equal division of household duties between spouses. Still other women tried to change the narrative by insisting that they were not left behind but chose to be single. These narratives, however, are predicated on the material and financial success of women. Therefore, they cannot be appropriated by *all* women. Furthermore, *shengnü* and *dangfu*—two powerful symbols—work together to stigmatize women. If you do not use dating apps, you risk becoming the former; if you use dating apps, you risk being called the latter.

Finally, in response to the state's denunciation, Momo underwent "sanitization" (T. Liu, 2016). Since then, Momo and other apps have strived to build a harassment-free space where women can report harassment to administrators. Violators are banished from the apps. This practice was appreciated by Jennifer, who rarely reported sexual harassment at her workplace. Dating app companies, in this regard, have acted as feminist allies, eradicating online sexual harassment and challenging the gender structure that has always favored men.

Although I applaud the dating apps for their harassment alerts and report systems, the contrast between this corporate initiative and the state's reluctance to combat offline harassment suggests that the state is shifting the responsibility to protect women onto companies. In 2016, a Guangzhou-based grassroots feminist group planned an anti–sexual harassment billboard campaign in the subway system (Lin, 2017). This could have been China's first subway advertisement to counter sexual harassment. However, after a year-long negotiation with the Guangzhou Administration for Industry and Commerce, the group's campaign was not approved and was never launched. The state's justified its refusal by claiming that the depiction of a human hand might cause public anxiety and asserting that grassroots organizations were not allowed to engage in public service advertising. This was hardly the only time the Chinese state had quashed efforts to reduce sexual harassment and gendered violence. The Chinese human rights activist Jinyan Zeng (2015) found numerous cases of violence against women in which local governments had not intervened. In December 2018, the Guangzhou Gender and Sexuality Education Center was also shut down, possibly due to its advocacy of women's rights (Feng, 2018). Together, these cases illustrate that in China, citizens' security is outsourced to companies by the state whenever the issue involved is deemed politically sensitive. The state's retreat from providing welfare to its citizens and the shifting of such duties to private companies are characteristics of neoliberalism (Harvey,

2005). When commercial entities such as Momo discipline online sexual harassers, women feel their voices are heard. However, the offline harassment of women remains untouched, unchallenged, and undisciplined by the state. Women can quit apps if they face online harassment; but they cannot quit their lives when harassment happens in their homes, workplaces, and neighborhoods.

The question of whether a dating app is a feminist technology is complicated by this contradictory evidence on identity, labels, and the changing gender structure. On the surface, dating apps seem to have led to more equitable gender relations. Based on Johnson's (2010) typology, these apps may be considered feminist. However, as Judy Wajcman (1991) suggests, technology can be used as a lens to view the larger culture. Through this perspective, it is apparent that dating apps are embedded in a larger sociopolitical environment where women are still subjected to surveillance with few legal protections.

So what does the future hold when a particular technology promotes more equitable gender relations only in a highly specific digital context while it is situated in an inequitable society in general? If we take the central tenant of technofeminism seriously—that technology shapes and is shaped by gender relations—we will expect one of three eventualities. The first possibility is that the liberation brought forth by a feminist technology in the specific digital context will have a spillover effect to the general society, which later becomes a favorable environment for more feminist technologies. The second possibility is that the hegemonic power inherent in the general society is so strong that it gradually takes away the disruptive potentials of the feminist technology, turning it into another conduit for patriarchy. The third possibility, which is theoretically possible but unrealistic, is the development of two separate spheres where gender relations are more equal in one sphere than the other.

I believe the first outcome is what feminist scholars and I would like to witness in the future. However, we cannot treat dating apps simply as a celebration of feminism as long as they conceal structural gender inequalities that have far-reaching ramifications. At best, dating apps can be liberating tools for women to exercise sexual agency, assert power over men through a feminine gaze, and be protected from sexual harassment.

At their worst, however, such apps hide the structural gender inequality embedded in the sexual double standard, marriage expectations, and state policies. To advance women's rights in contemporary China requires a fundamental change in the sociopolitical environment, not merely a technical solution.

CONCLUSION

The stories of my straight female informants have related the gendered experiences of using dating apps in China. These narratives have also demonstrated a characteristic of networked sexual publics: using dating apps is a nuanced experience that cannot be boiled down to a typology of motives. For example, although it is true that my informants used dating apps to look for sex, they discussed experimenting with love and sex, topics absent from their formal education. It is also true that they used dating apps to search for Mr. Right; however, such a desire was driven by the fear of being called *shengnü*. Attending to their personal circumstances and the Chinese social context allows us to more fully understand what dating apps mean to them.

Moreover, these female app users faced different challenges when using dating apps. Because dating apps are associated with *yuepao shenqi*, using them constitutes a taboo—one that applies more to women than to men. One of my informants deleted the app in front of her date. Others hid their dating app use or selectively disclosed to their friends and family how they used the apps. My informants also developed various methods to assess men's purposes, from judging the length of the men's written profiles to chatting with them. When they faced unsolicited sexual requests or sexual harassment, they engaged in one of four tactics—proactively indicating their revulsion against casual sex on their profiles, reactively reporting harassers to the app's administrators, ignoring harassing messages, or passively quitting the app. Although dating apps in certain ways have empowered women in China, once the larger sociopolitical environment is taken into consideration, it becomes apparent that the empowerment provided by the apps is bound by space: it does not extend beyond these apps. The everyday environment in China does not offer women equal respect or adequate protection.

Unlike cyberfeminism, which wishfully believes in the liberating potential of technology, a more grounded technofeminist perspective reminds us that technology alone cannot ameliorate a societal problem. Dating apps are not an antidote to gender inequality. Part of the reason is that it takes two to tango; dating apps are also used by men who compete for power. A way to assert dominance, then, is through the performance of masculinity. That is the focus of the next chapter.

3

CUTE IS THE NEW MANLY: PERFORMANCE OF CHINESE MASCULINITIES

Abandon any negative desire and don't give any excuse to myself. Just do it.

What I disagree is a majority of words of overgeneralization in the extreme.

I'd love to make friends from a diversity of positions, which can broden [*sic*] my horizon.

What's more, please help me for improving my English or any other problems from me, which is my pleasure.

This is a word-for-word quote from Victor's profile on Momo, which he wrote in English. The written profile appears to be sincere. It shows that Victor spent considerable time crafting it. It is not the longest I have ever seen on Momo, but it is definitely longer than most. More important, it was written in English, suggesting that Victor is *wenren* 文人 (an educated man). I asked Victor, "Your written profile is pretty long compared to others. Why did you do that?" He replied, "If I have to play, I have to play well. Some guys just write somethings very insincere. They meet [women] in person only for sex." I would not describe Victor as a playboy, but being thirty years old, good-looking, articulate, and currently in a dating relationship, he was by no means an innocent dating app user. He was not embarrassed at all to admit that he viewed Momo primarily as a hookup tool. Recalling his initial opinion of Momo before he installed it onto his phone, he said:

Momo was launched by positioning itself as *yuepao shenqi*. . . . Its promotional slogans were something about *yuepao shenqi*, "hot girls," "quick and easy." At that time, a flock of people said they'd like to try using the app. I did not know why women would use it, but I thought it may work, so I gave it a try.

In this quote, Victor interpreted Momo as a hookup tool. He also differentiated himself from women, who he suggested might not be excited by the promise of *yuepao shenqi* 约炮神器 (a magical tool for hookups; see chapter 1 for the term's etymology). Having used Momo for more than three years, he developed a methodology for soliciting sex from women on dating apps:

After we have a couple of casual chats, when the moment is right, when we are acquainted with each other, then I can try to test the water. I would say, "Let me tell you a joke" or "Yesterday, I heard a joke." A joke that is related to sex but not too vulgar. . . . Then you can see how the girl responds to sex-related topics.[1]

These three aspects of using dating apps—having a particular kind of self-presentation, holding a certain interpretation of dating apps, and devising an idiosyncratic belief and strategy for interacting with women—are not unique to Victor. In the previous chapter, I analyze the accounts given by straight female dating app users, illustrating half of the heterosexual dating app culture. In this chapter, I complete the picture by focusing on the experiences of straight male users. My analysis is founded on a Butlerian understanding of gender. Judith Butler (1999) argues that gender is not an innate attribute. Gender is *"a corporeal style,* an 'act' . . . which is both intentional and performative, where *'performative'* suggests a dramatic and contingent construction of meaning" (p. 177, emphasis in original). That is, there is nothing fixed or predetermined about one's gender: what makes one a "man" is the repetitive citation of a set of acts that are culturally associated with masculinities—what Douglas Schrock and Michael Schwalbe (2009) refer to as "manhood acts." There are no "men" if manhood acts are not carried out.

Ben Light (2013) argues that digitally networked media have reconfigured the production and reproduction of masculinities through digital affordances such as replicability, anonymity, and persistence. In this chapter, I explore how my straight male informants perform their masculinities through dating apps. Cultures from different times and places hold a different ideal of masculinities. Acts that are deemed masculine in some cultural contexts, contributing to the performance of masculinity, may not always produce the same effect in another culture. To address this need for cultural specificity, I introduce an indigenous Chinese concept of masculinities—the *wen-wu* 文武 (literary-military) dyad (Louie, 2002). With this literature as the backdrop, I tell the stories of my straight male informants. In addition to their interpretations of dating apps, I examine their self-presentations on dating apps and their interactions with women. I argue that my informants' gender performances reproduce existing gender inequalities and are complicit in maintaining them.

WESTERN MASCULINITY, CHINESE MASCULINITIES

Raewyn Connell's "hegemonic masculinity" (1987), an influential concept in the sociology of gender, describes the idealized forms of masculine behavior and appearance relative to the behavior and appearance of "women and subordinated masculinities" (p. 61). According to Connell, subordinated masculinities include gay, working-class, disabled, and Asian American. The most significant feature of hegemonic masculinity is heterosexuality. Behavioral manifestations of hegemonic masculinity have included positive practices, such as working hard and being a good father, and "toxic" conduct, such as physical violence against women and homophobia (Connell & Messerschmidt, 2005). Psychologists have also been interested in the consequences of men endorsing male role norms, a concept closely related to masculinity ideology. One of the most updated typologies of these male norms includes avoidance of femininity, negativity toward sexual minorities, importance of sex, self-reliance, toughness, dominance, and restrictive emotionality (Levant, Hall, & Rankin, 2013).

The concept of "hegemonic masculinity," however, has been criticized.[2] The critique most relevant to my analysis is that the conceptualization of hegemonic masculinity is based on a Western ideal of masculinity. Connell later recognizes this limitation and suggests that attention be paid to the international geography of masculinities (Connell & Messerschmidt,

2005).[3] In analyzing the performance of masculinity on dating apps in China, we must consider how the performative acts carried out locally through these apps reflect the regional cultural ideal of Chinese masculinities and generate dialogue through the global circulation of masculine images.

What does it mean to be masculine in the Chinese context? A common way of conceiving Chinese masculinity and femininity is to resort to the *yin-yang* 阴阳 dialectic, symbolizing the two opposite yet interdependent forces of nature contemplated in Daoism. However, building on both ancient texts and contemporary popular culture, Kam Louie (2002) proposes the *wen-wu* dyad to better understand Chinese masculinities. *Wen* refers to cultural attainment; *wu* represents physical prowess. He maintains that the *yin-yang* dialectic is less suited to masculinities because both *yin* and *yang* can be applied to men and women; *wen* and *wu* are more appropriate because they are indigenous concepts that exclusively apply to men in the Chinese culture. The archetypal figure of *wen* masculinity is Confucius, whereas *wu* masculinity is exemplified by a famous general from the late Eastern Han dynasty named Guan Yu. Louie elaborates:

> *Wen* is generally understood to refer to those genteel, refined qualities that were associated with literary and artistic pursuits of the classical scholars. . . . *Wu* is . . . a concept which embodies the power of military strength but also the wisdom to know when and when not to deploy it. (p. 14)

Wen and *wu* are the qualities a Chinese man wants to have. However, they are anything but equal. In many Chinese historical periods, *wen* was considered superior to *wu*. In a sense, *wen* can be considered a "more elite masculine form" than *wu* (p. 18). Further, sexual dominance over women, often associated with the image of muscular men in the West, is a property of *wen* masculinity, not *wu* masculinity. Historically, it was considered socially acceptable for a *wenren*, a man with *wen* qualities, to seek out women and reject them if they were inconvenient. A *wu* hero, on the other hand, "must contain his sexual and romantic desires" (p. 19).

Cultural ideals evolve, and so do Chinese masculinities. China's economic reform since the 1980s has reshaped the *wen* ideal. Some management theorists have been eager to incorporate the teachings of Confucius into business practices. As Louie (2015) notes, this gradual evolution has rendered business something that a *wenren* can legitimately pursue. As a result, *wen* has been reconfigured to include economic success in addition to cultural attainment. Regardless of whether masculinity is *wen* or *wu*, it must be earned. Derek Hird (2016) investigates the interplay between masculinity and class. His main informants, white-collar Chinese men living in Beijing, framed and reinforced their gendered and class privileges through the language of freedom, choice, and equality. The pursuit of financial success as a performance of masculinity is also observed by Fengshu Liu (2019) among secondary school boys. These boys perceived financial success to be the prerequisite of the "three goods"—a good life, a good person, and most important, a good man.

Economic success, however, is not always attainable. This has been particularly true for rural-to-urban migrant men. Susanne Yuk-Ping Choi and Yinni Peng (2016) coined the term "respectable manhood" to describe the way Chinese men who negotiated the meaning of manhood when moving from rural villages to cities for work. Respectable manhood, they write, "resists the colonization of private life by money, and views money as being too often a source of evil and family ruin" (p. 118). Due to their low financial status and skills, these men rejected the definition of manhood as economic success or professionalism. Instead, they emphasized their contributions to domestic chores and childcare.

From children to adults, we can see the sweeping influence of financialization on *wen* masculinity in China. *Wen* masculinity evolved into a concept similar to what Raewyn Connell and Julian Wood (2005) call "transnational business masculinity." However, there is a fundamental difference between the two. In exploring transnational business masculinity among Australian managers, Connell and Wood point out that self-doubt and insecurity engendered by the global economy had brought more challenges to managers performing their masculinity. In contrast, the financialization of *wen* masculinity in China suggests that today there is another way to achieve *wen* masculinity apart from being a genteel *wenren*. If the former is narrowing the potential to perform masculinity, the latter arguably opens up more potential.

This indigenous understanding of masculinity in the Chinese context—the idea of *wen-wu* masculinities—allows us to interpret the gender performance of Chinese men more accurately. For example, the mastery of music and poetry that may not be commonly regarded as masculine in the West is indicative of *wen* masculinity. Keeping this in mind is important because some of the men's gender performance on dating apps, which I reveal next, may not appear to be masculine according to Western hegemonic masculinity.

DATING APPS AS THE STAGE FOR MASCULINITY

Researchers of media studies, cultural studies, and public health have examined different phases in the use of dating apps. Colin Fitzpatrick and Jeremy Birnholtz (2018) identify three phases of interaction in the use of Grindr—profile creating and viewing, chatting online, and meeting in person. The first phase is analytically distinct from the second and third phases because it does not involve interactions with a person. The second and third phases differ only in terms of where the interactions take place. My analysis of the narratives given by my straight male dating app users suggests a pre-interactive phase—how men interpret dating apps. Users' interpretations indeed influence how they create their profiles and how they interact with others on the apps. Writing this new phase into my analysis, I categorize my informants' performance of masculinity according to the following phases—interpretation, self-presentation, and interaction.

INTERPRETING DATING APPS: SEX, LOVE, AND WORK

In the previous chapter, I show that my straight female informants had four ways to interpret dating apps—as a laboratory for sexual experiments, a springboard to romance and marriage, a third place between home and the workplace, and a gateway to new worlds. What sort of interpretations did my straight male informants have of dating apps? Three dominant interpretations emerged. These interpretations are not mutually exclusive: a single user can hold all of them. First, similar to their female counterparts, male informants used dating apps such as Momo and Tantan to seek casual sex. However, there was a significant difference in how they interpreted "seeking sex." For my female informants, seeking sex on dating apps allowed them to develop their sexual agency and their thoughts on love and sex. My male informants, however, often framed sex seeking on dating apps as "physiological":

> The reason I downloaded [Tantan] was to hook up. I just finished high school and was very bored. I was stuck at home, got nothing much to do, so I downloaded it. . . . I do not know about others, but for my current situation, it is the kind of physiological dependence. Everybody has this thought, at least for a man. He uses the app because of his need.

These are the words of Xiaoli, age nineteen, a college student. His use of phrases such as "physiological dependence" and "everybody has this thought" not only legitimates his own use of Tantan for hookups by linking his behavior to a biological mechanism and social norm but also is reminiscent of Abraham Maslow's (1943) hierarchy of needs. Young men at his age regard hookups as a fulfillment of their physiological need—the lowest level on the hierarchy—whereas my female informants viewed hookups as a journey of self-actualization—a need at the highest level. Victor, mentioned at the beginning of this chapter, compared sex with fast food:

> If I am at home, I can only eat home-made food. But if I dine out, I can eat anything. I don't mind there is a new *cha chaan teng* or coffee shop.[4] If I like it, I will dine there. It's like a shopping mall. I am holding a shopping mall in my phone.

Victor's narrative analogizes hookups with the fast-food culture that is grounded in convenience and choice. This is a sharp contrast to the deeply personal reflections on hookup experiences discussed by my female informants.

Another difference in the way my straight male and female informants interpreted dating apps was how they viewed the search for a romantic partner or a spouse. My single female informants, such as Jessica and Nancy who were in their late twenties, were worried that they would be called *shengnü* 剩女 (leftover women). To them, dating apps were a springboard to eliminating this potential stigma. For my male informants of a similar age, their use of dating apps for romance did not take such a serious tone. For example, Jiazhi was twenty-eight years old when we met. He had received his undergraduate degree in Japan and worked as an office executive at a Japanese company in Guangzhou. He was single and planned to marry at around the age of thirty-five. He logged onto Tantan several times a day, but he said, "It is better to get married after the career, social experiences, and personal relationships are more mature." Dylan, age twenty-nine, an electronic developer and entrepreneur, although sensing some pressure from his family to get married, was capable of ignoring it. He said:

> If I do not have a career, I have no intention of dating. *(Researcher: Is there any pressure from your family?)* I should say that there is pressure, but I am not stressed because I am relatively strong. Not only am I the only one who has gone to a college in my family; I am the only one who has done so in my entire extended family. . . . It has almost one hundred people and I am the only one who finished the college. So I often dominate the conversation.

Jiazhi's prioritization of his career and Dylan's undergraduate degree gave them the leverage to eschew the pressure from their families to marry, even though both were approaching thirty. This suggests an unpleasant irony: for a woman, a career and a college degree are liabilities, placing her at the risk of being *shengnü*. It is little wonder that my straight female informants viewed dating apps as more important to their romantic lives.

The third difference between my straight male and female informants was the novel finding among my male informants that they associated dating apps with business. Dylan, whom I just introduced, for example, came to Guangzhou at the beginning of 2016 to launch a business. He said he wanted to meet business partners on Momo and Tantan. When I asked him if he connected with men on these apps, he responded that he mainly connected with women: "I look at their profiles, see if they run their own businesses or are in the process of starting up businesses. I add those who show a sense of entrepreneurialism, talk to them, learn from them." Thus, Dylan framed his reaching out to women on dating apps as business-related. He eventually found a female business partner on Tantan to sell red wine with.

Clement, age twenty-eight, shared with me another business-related use of dating apps that I was surprised by. He had gotten to know a couple of female sex workers through these apps. They sometimes acted as his "cover" during "corporate entertainment" that included nightclubs and prostitution.[5] The following was his elaboration:

You can only get to know girls in this profession through apps. . . . During my last project as a financial consultant, my clients arranged a nightclub event. Sex services are usually paid for in advance by clients. If I have to pick a woman, I prefer someone I am familiar with instead of a random woman from the club. . . . When she [one of his female sex worker connections] was there, the client thought I had a regular playmate so did not assign me a random woman.

It is highly doubtful that Clement was seeking this sort of arrangement when he began using dating apps. As an interview setting can be a scenario in which male informants seek to perform their masculinity, Clement's narrative could have been a performance of masculinity for me as a researcher (Schwalbe & Wolkomir, 2003; see appendix for a more detailed reflection). However, out of the potentially unlimited ways of framing his own experience, he chose to frame his use of apps as an occupational convenience. This suggests that "business" occupied a significant position in his dating app use.

These interpretations of dating apps as related to business illustrate the interpretive flexibility of technology users (Pinch & Bijker, 1987). When Momo was launched in 2011, its marketing materials highlighted its use for developing sexual relationships with strangers (T. Liu, 2016; as Victor described). Although Momo's cofounder Tang Yan made a public statement in 2014 where he acknowledged that some Momo users used the app for business (Y. Tang, 2014), the app's design has never catered to business users. Tantan has largely followed the interface of Tinder. No official claims have ever been made that it was designed for business. So using Momo and Tantan for business demonstrates users' agency.

CREATING PROFILES: GENTLE, INNOCENT, AND FINANCIALLY SUCCESSFUL

Dating apps afford visibility. When strangers meet on dating apps, their initial impressions are built entirely on their profiles (Ward, 2017). Shaka McGlotten (2013) refers to the crafting of a profile and the effort going into making the best photograph—"good grooming practices, trips to the gym, and carefully selected and rehearsed readymade examples of who you 'really' are" (p. 128)—as image labor. Yet Chris Haywood's (2018) recent research on young British men's use of Tinder found that these young men often denied that they spent effort on their photographs. Haywood calls this "effortless achievement"—a rhetorical tactic the young men used to protect themselves from potential failure. Using this tactic, the young men explained their failure to attract women on Tinder by their poor choice of photographs, not their lack of ability to attract women. "The threat to their masculinity is," he summarizes, "deflected through an effortless achievement" (p. 152).

I did not hear the rhetoric of "effortless achievement" from my male informants. Instead, most of them freely told me about their image labor and the ways they picked photographs they believed would impress female dating app users. Photographs convey one's "sexual capital," which Adam Green (2014) defines as "the degree of power an individual . . . holds within a sexual field on the basis of collective assessments of attractiveness and sex appeal" (p. 48). Spaces such as bars, fitness facilities, public bathrooms, and dating apps have unique configurations of members and thus constitute different sexual fields. Via members' ratings of each other and via sexual socialization, a standard of desirability gradually emerges within a sexual field. Individuals possess what Green describes as "currencies of sexual capital," and these currencies are evaluated against the standard of desirability in a given sexual field.

What kinds of photographs did my informants use to enhance their sexual capital on dating apps? Performing Western hegemonic masculinity would require demonstrations of bodily strength through displaying postworkout images or images of muscular bodies (Hakim, 2018; Miller, 2015a). None of my straight male informants did this. Instead, three alternative types of profile photographs were common among my informants.

The first type was images of animals. My male informants said that some women found their profiles more attractive based on a shared love of animals. For example, Eric, age thirty-one, expected women to perceive his keeping pets as an indicator of his gentleness. Similarly, Clement displayed an image of his cat as his main photograph, with his own photographs visible only after clicking to access his profile. He explained, "It is strategic. Pets may help to attract women." John, age thirty-six, never used his own photographs on Tantan. Instead, he displayed a cartoon drawing of several cats. The use of animal photographs should not be seen as an isolated incident. In Western popular culture, the photographic subgenre of "guys and animals" emerged in 2010. On Instagram, one can search for #tinderguyswithcuteanimals or #hotguysandbabyanimals. The 2019 Australian Firefighters Calendar, featuring typically muscular firefighters and adorable animals, has also captured attention from the worldwide tabloid news media and social media.

Closely related to these heartwarming photographs is a second type of photograph that my informants called *maimengzhao* 卖萌照. *Zhao* is a photograph. In the Chinese digital context, *maimeng* (selling sprouts) means "acting cute." The people in the photographs position themselves to hint at their innocence, for example, by doing a V-sign or touching their cheek with their hand. Such photographs are usually brightened, with a smoothing effect applied to the skin. Photographs of this kind, heavily influenced by Korean pop culture, appear both online and offline in China (figures 3.1 and 3.2), defining what is "good-looking" for a man. This type of self-presentation resembles the ritualization of subordination identified by Erving Goffman (1976) in women's pictures in magazines. However, the ritualization I am referring to here is particularly performed by younger men. In his profile photograph on Tantan, Nathan, age twenty-five, covered his mouth with his fingers. He said he imitated

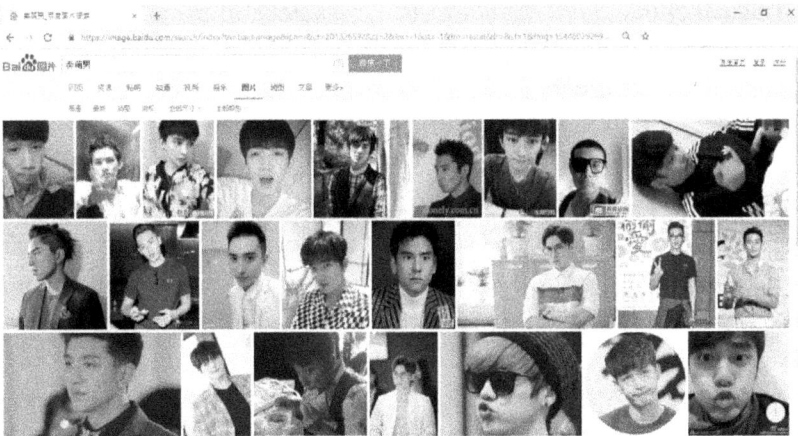

Figure 3.1
Results of a search for *maimeng nan* 卖萌男 (cute-acting men) on Baidu, a major Chinese search engine. (Screenshot taken by the author on December 14, 2018)

Figure 3.2
A billboard advertisement in a Guangzhou metro station. The skin of the male celebrity appears to have been lightened and smoothened using editing software. (Photograph taken by the author on July 31, 2017)

the typical gesture of Korean pop stars. The main photograph displayed by Xiaolong, age twenty, was similar. When I asked why he chose this particular photograph, Xiaolong's reply was straightforward: "I only post pictures that I think are especially good-looking."

Displaying photographs of one's lovely pets or *maimengzhao* would hardly be considered masculine in the Western hegemonic ideal. In these genres of photographs, men show their childlike, innocent side and rarely demonstrate their physical strength. These images can even be regarded as an appropriation of the Chinese "cuteness" culture among urban female youth (Z. Qiu, 2013). However, if we treat these photographs as a demonstration of gentleness and refinement, these self-presentations are examples of performing *wen* masculinity.

The third type of profile photograph flaunts one's financial success. Older men tend to include one or two photographs that subtly hint at their financial status. Fung, age thirty-six, owned a coffee shop. Although he did not explicitly say he was the owner of the shop on his profile, he displayed photographs of the place and wrote "Drop by our coffee shop any time" in the text of his profile. Dylan, who said he used Momo and Tantan to look for business partners, put up three photographs of himself—in a business suit, at a boat party, and on a nice balcony of a building that could be mistaken for his apartment. He said, "I want to let people see my life, my business. . . . *(Researcher: Is that balcony in your house?)* No, I do not own a house. That photograph was taken on a hotel balcony." In this example, Dylan carried out his "image labor." The difference between his labor and what McGlotten describes is that instead of going to the gym to buff up, he photographed himself in a hotel room, creating an impression that he owns a luxury home.

This type of self-presentation, together with the business-related interpretation of dating apps I discussed previously, reflects the financialization of *wen* masculinity in recent decades. As Louie (2015) notes, China's recent rise as a global economic powerhouse and the marketization of Chinese society have recently recast *wen* masculinity as financial success. Therefore, interpreting dating apps as a platform related to work or to highlight one's financial status with photographs are some ways my informants performed their *wen* masculinity. In fact, none of my straight male informants, regardless of their background, focused on presenting their *wu* masculinity.

INTERACTING ON AND BEYOND DATING APPS: MAXIMIZATION OF OPPORTUNITIES

Moving to the third phase of dating app use—interaction with female users—I noticed three areas my informants worked on to maximize their dating or hookup opportunities. The first area was the initial contact strategy. Dating apps afford access to a large pool of potential partners that a user is otherwise unlikely to meet. However, this can create the problem of choice overload for some users.[6] My male informants developed their own contact strategy to manage the vast number of choices. The two initial contact strategies most often used by my informants were *taking-all* and *filtering*. Taking-all is like casting a net to catch fish and is used by those who do not find having an abundance of options suffocating. On Tantan, which functions like Tinder, some of my informants swiped right on nearly every woman they saw on the app and waited for a mutual match. Starting with such a large number of right swipes guaranteed at least a handful of matches. On Momo, where mutual liking is not a prerequisite for conversation, this strategy manifested as users sending a "Hi" to almost every woman encountered on the app.

Users can also regulate choice overload by using the filtering function to reduce the number of potential partners (L. S. Chan, 2018a). Many informants, like Clement, set an age range to exclude women who were older than they were. Some used location as a filtering tool. Roy, age twenty-one, told me that whenever he visited a bar with his male friends, he checked out Momo to see if there were any girls nearby he was interested in. "If I am there and she is also there, the chance [of her coming over to my table] is pretty high." He estimated that his success rate was around 70 percent when he reached out to women who were nearby. Therefore, location and distance provided a heuristic for Roy to determine whom he would send a message to. In this way, the affordance of proximity was also crucial to those soliciting hookups. Fred, age twenty-five, was very concerned about the distance between him and the women he wanted to have sex with. He explained:

> I log on to the app, set a center, and search outward. If a woman is more than four kilometers away, I won't contact her, no matter how attractive she is. . . . Sometimes I say, "We are so close, you should come over." Women who believe in fate tend to agree to visit.

The second area concerns the order of messaging. After users narrow down the number of potential partners, they may start a conversation with other users. Dating convention in the West gives more control to men, who are expected to lead the courtship process (Bailey, 1988). More than half of my informants followed this cultural script mandating that men make the first contact with women. Some of the informants said that being proactive gave them an advantage. Eric remarked that women preferred proactive men to passive men. Anthony, age twenty-eight, perceiving women as more reserved, told me with confidence: "If you don't contact her first to give her a good impression of yourself, a conversation may never happen."

The third and final area in which my informants sought to maximize their dating or hookup opportunities was the categorization of women. Before I elaborate on this issue, I would like to reflect on my role as a male researcher in this project. I have pointed out that, in the introductory chapter, narratives I gathered from my informants were coconstructed. By coconstructed, I mean that what my informants had told me, to a large extent, depended on how I interacted with them. In the reflexive account of his research of male pornography viewers, Florian Vörös (2015) describes himself being in hegemonic complicity. That is, he had to bracket out his feminist judgment during the interviews in order to make his male informants comfortable in sharing their pleasures and fantasies around pornography with him. Looking back, I realize that in my research I also engaged in hegemonic complicity; whenever my male informants described practices that I deemed sexist, I held my tongue.

My informants assessed the probability of developing a successful relationship with each woman and focused their time and energy only on promising candidates. This helped them avoid choice overload. For example, in his worldview, Victor put women into four groups: soulmates, sex partners, work partners, and "miscellaneous." The last group was women who had the potential to become soulmates. He did not think that soulmates and work partners existed on dating apps, but he often encountered sex partners and "miscellaneous" women on Momo. Because "miscellaneous" women could become soulmates, according to his system, he added these women to his WeChat, where he maintained contact with his everyday friends. He engaged in more personal disclosures with this group of women. However, the women he considered sex partners remained only on Momo. About this group of women, he said, "It's not necessary to talk too much; the face-to-face meeting is just for sex." Because two different interactive styles applied to these two groups of women, segregating them into two platforms—WeChat and Momo— helped Victor manage his communication.

Fred's classification was more complicated. He used a two-step classification scheme to categorize the women he met on Momo. The first step was to assign attractiveness. He said women who were extremely beautiful were out of his league, so he would not waste time talking to them. He approached only the remaining women, soliciting their WeChat contacts. He described transfer of these women from Momo to WeChat as "the centralized management of the back palace." In imperial China, the back palace was where the emperor's concubines lived. The second step of his classification scheme took place on WeChat, where he assigned a label to each of the women. He explained, "The letter B means that I have met them face-to-face before. The letter F is given to those I have not yet met. The letter Z basically denotes those whom I do not want to meet." To avoid mistaking the women he met on Momo for the women he met on everyday social occasions, he assigned the letter A to the latter group. Based on the category a woman was in, his contact frequency changed. He worked harder on group F than group B because trust between him and the women in group F had not yet been developed.

REPRODUCING GENDER INEQUALITY

Up to this point, I have described how my male informants performed their masculinity on dating apps. I have introduced the notion of Chinese masculinities, particularly the *wen* component, and argued that it provides a more accurate lens to analyze men's gender performance in the Chinese context. Some men performed their masculinity through the use of animal photographs and *maimengzhao*. I have also discussed the financialization of *wen* masculinity.

Given these variations in masculinities, can we say that Chinese masculinities have reproduced gender inequality in China? This is an empirical question because some versions of masculinities—reluctantly or willingly—are more open to equality with women. For instance, Choi and Peng's (2016) research shows that Chinese rural-to-urban migrant men partially forfeited their dominance by compromising on their marital power and division of labor within the household. Michael Messner, Max Greenberg, and Tal Peretz (2015) document the experiences of three generations of men, the majority of whom are straight, who actively participated in the women's movement in the United States. In 2018, the world also witnessed male celebrities, mainly in Western countries, voicing their support for the #MeToo movement against sexual harassment and violence. As Connell and Messerschmidt (2005) put it, "the conceptualization of hegemonic masculinity should explicitly acknowledge the possibility of democratizing gender relations, of abolishing power differentials, not just of reproducing hierarchy" (p. 853).

Given the plurality of masculinities, Schrock and Schwalbe (2009) remind us to focus on how "manhood acts" create inequality. Manhood acts enable men to "distinguish themselves from females/women and thus establish their eligibility for gender-based privilege" (p. 287). For instance, some manhood acts involve occupational norms, such as working long hours, enduring pain, and celebrating extreme rationality. These actions discourage women from participating in certain occupations (Cohn, 1987; Cooper, 2000; Curry, 1993).

The narratives of my male informants revealed how they sustained their gender-based privilege. First, informants like Dylan, who used dating apps for business and curated images of financial success on their profiles, were reasserting the age-old stereotypical relationship between masculinity and work: men should focus on their careers, should be ambitious, and should be the breadwinners for their families. My informants also encountered women who used dating apps for business purposes. However, when they talked about these women, they expressed disdain. According to my informants, there were three kinds of women who actively used dating apps for business-related purposes. The first were saleswomen who usually promoted financial products and gym training sessions. These women were often very upfront about their business purposes. My informants said they had no objection to the presence of these women on dating apps as long as they did not keep sending promotional messages to them. The second kind comprised sex workers. My informants said these women usually left their profiles blank. As these women's accounts often garnered complaints and were banned, they invited male users to connect via WeChat as soon as possible. My informants described these women as annoying because they often sent pornographic photographs to them. The third type of women were bar promotors, notoriously known as *jiutuo* 酒托 (alcohol support). My informants despised them the most because they pretended to be ordinary female users and lured male users to visit expensive nightclubs or bars. However, my informants said they had ways to identify *jiutuo*. Bob, age thirty-seven, said, "If she has opened her account within the last month and initiates a conversation with you, then asks you to add her WeChat, she is likely to be problematic—whether *jiutuo* or a trickster." Fred agreed and added that *jiutuo* were always extremely particular about where to meet. "If you suggest another place, she will lose interest in you immediately."

My informants clearly differentiated "good" business on dating apps, mainly conducted by men, and "bad" business, solely carried out by women. One could argue that because *jiutuo* make a profit by cheating and sex workers cause annoyance, their businesses should not be allowed on dating apps and it should be legitimate for men to criticize them. However, this argument ignores the gendered labor market and the massive migrant flow in China, which already disproportionately favor men and have left fewer chances for women to pursue otherwise socially respectable and sustainable careers (C. C. Fan, 2003; Gaetano, 2015). Some women work as sex workers or *jiutuo* because they cannot find other jobs that can sustain their life living in a city. Therefore, when men make the distinction between "good" and "bad" businesses on dating apps, they are claiming their second round of gender-based privilege, after their first round of privilege in the labor market.

Second, researchers have found that some men in college collectively objectify women to mutually affirm each other's manhood (Bird, 1996; Martin & Hummer, 1989).[7] Haywood (2018) reports that his young British male informants gather in groups to look at Tinder and evaluate the women showing up on the app. Although none of my straight male informants told me they did this, Victor and Fred's quantifying attractiveness epitomizes the objectification of women. Assigning grades to women based on their attractiveness is not uncommon. Part of the huge entertainment industry has been dedicated to publicly ranking women. However, with the help of communication technologies, the level of objectification has intensified and become more personal. Victor used WeChat to maintain relationships with "soulmates," while retaining his sex partners on Momo. Fred used WeChat basically for "bookkeeping." In his narrative, he even used the phrase "the back palace" to describe the women he met on Momo, as if they were his personal concubines. His grading of women based on their attractiveness and his sole pursuit of women's bodies for sex are typical of the "seduction community" in the United Kingdom or the "pickup community" in the United States (O'Neill, 2018). These communities consist predominantly of men who wish to learn how to gain women's trust and, thereby, sexual access. To some degree, this "toxic" aspect of masculinity is implicitly encouraged in the Chinese cultural ideal of *wen* masculinity. I mentioned that *wenren* have often been associated with sex and, according to traditional Confucian beliefs, women are men's property. When the design of dating apps packages everyone as a profile, women become generic products to be catalogued by male dating app users like Fred and Victor. This outcome results not directly from the design of the dating apps but from the men's predatory interpretation of the dating apps.

What about the lovely photographs of cats or the cute-acting portraits men use on dating apps? The latter type of photograph may be considered as a performance of "metrosexuality" (Simpson, 1999), the metropolitan lifestyle where straight men pay meticulous attention to their hair, skin, bodies, and fashion—activities traditionally associated with femininity. Do these acts also contribute to the perpetuation of gender inequality? Probably not—at least not directly. However, these visual forms are similar to Tristan Bridges's (2014) discussion of "hybrid masculinities." Hybrid masculinity accounts for how urban, heterosexual, white men incorporate "gay aesthetics" to create an alternative form of masculinity that is distinct from the "tough" and "toxic" kind of hegemonic masculinity. Bridges argues that although such incorporation indicated straight men's acceptance of homosexuality, they still maintained their associated heterosexual privilege. He writes, "by casually framing being gay only as fun and exciting, this practice allows these straight men to ignore the persistence of extreme sexual inequality and the hardships that actual gay men face every day" (p. 79). The softening of masculinity using animal photographs and *maimengzhao*, one that appropriates the "cuteness" conventionally performed by women in East Asia (Abidin, 2016; Z. Qiu, 2013), represents an expansion of the styles that men can choose from. However, they are one-way, reinforcing the gender-based privilege men enjoy without disrupting the existing gender hierarchy.

CONCLUSION

As Connell and Messerschmidt (2005) write, "a regional hegemonic masculinity . . . provides a cultural framework that may be materialized in daily practices and interactions" (p. 850). In this chapter, informed by this idea of regional masculinities, I have drawn on Louie's theorization of Chinese masculinities. Chinese masculinities comprise *wen* masculinity, which emphasizes cultural attainment and sociability with women, and *wu* masculinity, which involves physical strength and sexual abstinence. *Wen* masculinity has recently undergone an economic transformation, increasingly associated with financial success. The regional understanding of Chinese masculinities provides a more culturally sensitive framework for analyzing gender performances of my straight male informants on dating apps. I have also considered how the self-presentations of my informants fit into the global circulation of "good-looking" images from Korean pop culture.

I have identified three phases of app use in which men perform their gender and have discussed how these performances reproduce gender inequality. First, my informants interpreted dating apps not only as a platform to fulfill their physiological needs but also as a platform to develop their business. Their approval of "good" business and critique of "bad" business often carried out by women ignored the gendered labor market that acts unfavorably against women in the first place. Second, on their profiles, some of the men presented their gentleness by posting animal photographs. Others demonstrated their cuteness or hinted at their financial success. Such performances may not be considered masculine in Western hegemonic masculinity. However, within the framework of Chinese *wen* masculinity, men who post animal photographs, such as Clement and John, and those who display *maimengzhao*, like Xiaolong, are no less masculine than users who highlight their muscular or athletic bodies. Finally, the informants devised different strategies to handle the vast number of potential partners they met on dating apps. Some connected with all women, while others were more selective. My description of Victor's and Fred's classification schemes demonstrated an intensified version of the objectification of women. For these reasons, I have warned against regarding a softening of masculinity as progress in gender equality.

My analysis in this chapter, together with chapter 2, suggests that contrary to what cyberfeminism has envisioned (see chapter 1), the space of dating apps is not completely liberating for women. If there is room for feminist resistance in networked sexual publics, there is also the threat of redomination. Dating apps concomitantly allow men to maintain their power and fail to fulfill the feminist hope for more equal gender relations. It seems to me that if men cannot learn to conceive of gender dynamics beyond a zero-sum game, they will see the empowerment of women as a threat and women's gains inevitably as their losses. In other words, if men cannot learn to share power, the global and regional uprisings over the women's rights movement and the emergence of new technologies will continue to touch a nerve.

In the next two chapters, I turn to a related set of questions: In what ways are dating apps related to current queer politics? What do gay and lesbian dating apps mean to their users? And how does the use of dating apps create a more livable space for them?

4

CYCLES OF UNINSTALLING AND REINSTALLING: CONTRADICTORY AFFECTS IN GAY APP USE

I have uninstalled [the app] multiple times. It is a cycle. I believe many Blued users have a similar experience to mine. For this app, after you uninstall it, you install again. Then [you] feel it is meaningless and uninstall it again. For different reasons or feeling upset, you just don't want to log onto the app. Then you re-install it the next day.

I quietly listened to River, age twenty-six and gay-identifying,[1] describe his entanglement with Blued, a dating app that he had been using for six years. His repetitive uninstalling and installing of a dating app for reasons unrelated to finding a romantic partner was a rather distinctive phenomenon observed primarily among my queer male informants.[2] In this chapter, I explore this phenomenon through the lens of affect. My arguments are twofold. First, I argue that this phenomenon is a manifestation of the ambivalence toward gay dating apps felt by my queer male informants. Second, I also suggest that these contradictory emotions emerged from both using these apps and living as gay men in China.

A plethora of studies have been conducted on why we adopt or abandon a specific communication technology. For instance, the technology acceptance model (TAM) (F. Davis, 1989), a major theory in the field, explains why we adopt particular information systems in an organization. According to this model, the perceived usefulness and ease of use of a system are two major predictors of the intention to use it. This model and its variations have received robust support in empirical studies out-side organizational contexts, examining, for example, the adoption of social media (e.g., Choi & Chung, 2013; Dutot, 2014; Rauniar, Rawski, Yang, & Johnson, 2014). Meanwhile, research on abandoning a digital medium has been more limited. Based on the nascent scholarship on nonuse and abandonment (e.g., Birnholtz, 2010; Mainwaring, Chang, & Anderson, 2004; Portwood-Stacer, 2013), Jed Brubaker, Mike Ananny, and Kate Crawford (2016) explore why people quit Grindr. The reasons they have identified include the following: (1) users found that using the app was time-consuming, (2) they were unable to look for the relationships they wanted on the app, (3) similar profiles always showed up, and (4) the app promoted the culture of "always keep looking" (p. 380). This analysis might very well explain why people disconnect from a dating app. However, it does not show why people reconnect again in the cycles that characterized queer Chinese men's use of dating apps. In his study on gay men's use of an online dating website in Australia, Elija Cassidy (2018) proposes the useful idea of "participatory reluctance" to capture the constant tug of war between use and nonuse, what he defines as "par-ticipation in a state of discontent—neither fully active or absent" (p. 7). Although his informants recognized that the relationships they wanted could not be fulfilled on the site, they found no alternative to socialize with other gay men. Therefore, they had no choice but to continue using the website.

The explanations above share two motifs. All authors appeal to users' cognitive appraisal of the usefulness or rewards derived from using a spe-cific technology. In addition, they use in-app experiences to explain for connections and disconnections. Elaborating on his "media go-along" methodology, Kristian Jørgensen (2016) points out that "both naviga-tion and its associated feelings constitute the media environment as a place for the user" (pp. 38–39) in the dating app context. In other words, users emotionally engage with the media environment through feelings, which help them navigate and make sense of it. By turning to affect in this chapter, I move away from the cognitive paradigm exemplified by TAM. The option I explore is to treat dating apps as an "affective fabric" (Kuntsman, 2012) interwoven with various kinds of emotional threads across multiple contexts. Specifically, I argue that the ambivalence my queer male informants felt toward the dating apps came from their in-app experiences and that such ambivalence was also deeply rooted in the queer politics of China. I call this second type of emotions "out-of-app emotions," as opposed to "in-app emotions," which are triggered by the direct use of the apps.

To support my argument, I first offer an abridged history of male homosexuality and some data on gay dating app usage in China. This background information is needed to understand the affective dimension of queer politics in China, which is very different from that in the United States. Since the mid-1990s, social sciences and humanities scholars have turned their attention to affect and emotions. I differentiate two differ-ent "schools" of affect theory (Schaefer, 2015). The first school considers affect and emotion as two distinct registers, whereas the second uses the words *affect* and *emotion* interchangeably. My thought aligns more with the second school. Specifically, I draw on the theories of Sara Ahmed (2004b, 2010) about the intricate relationship between emotions, objects, and experiences. Using her concepts, I analyze the source of ambivalence reported by my queer male informants in relation to dating apps and discuss why such ambivalence was not prevalent among my straight informants.

LIFE AS A GAY MAN IN URBAN CHINA

Although descriptions of male same-sex practices are not uncommon in Chinese historical records, prior to the twentieth century traditional Chinese culture did not have a specific term for homosexuality (Chou, 2002).[3] It was not until the 1910s and 1920s, when medical journals, magazines, and sex education manuals from the United Kingdom and Germany were translated into Chinese, that the term *homosexuality* was introduced into the Chinese lexicon (Sang, 2003).[4] From that time, writ-ers began using pejorative terms such as *pi* 癖 (obsession) and *renyao* 人妖 (human monster) to refer to men who have sex with men (Kang, 2009).[5] Coinciding with this development, the first half of the twentieth century saw the downfall of the nation, due in large part to the threat of Japanese imperialist expansion. In these turbulent times, a generation of cultural conservatives and tabloid writers began to treat male homosexuality "as a cause of moral confusion, a symptom of political corruption, a social vice, a crime, a sign of colonial oppression and national humiliation, and a behavior alien to the Chinese" (p. 86). These writers began to pathologize homosexuality as a dire threat to national identity.

When the Communist Party of China (CPC) founded modern China in 1949, no law explicitly criminalized male homosexuality. However, if a man was found engaging in same-sex activities, he would be sanctioned by the party or sent to a labor camp (Davis & Friedman, 2014). In 1978, male homosexuality was officially outlawed. Falling under the vaguely defined crime of "hooliganism," male same-sex activities were codified as a cause of public disorder. According to Yinhe Li (2014), the first mention of homosexuality in the mass media appeared in *People's Daily*, the mouthpiece of the CPC, in 1980. The newspaper linked homosexuality with the fall of spirituality and moral degradation in the West. In 1986, the first AIDS case was reported in China, which the media framed as the consequence of living a Western liberal lifestyle. Thereafter, the second edition of the *Chinese Classification of Mental Disorders*, published in 1989, classified homosexuality as a mental disorder. In a nutshell, from the 1970s to the 1990s, being gay in China went from being regarded as immoral to being illegal and even a mental disorder.

The government and medical establishment changed their stance toward homosexuality around the turn of the twenty-first century, which shifted how gay men perceived themselves. In 1997, the law against male homosexuality was abolished. In 2001, the *Chinese Classification of Mental Disorders* removed homosexuality as a mental illness. During this time, a disparate "structure of feelings" (Williams, 1961) between the older and younger generations also emerged. Based on his extensive interviews with gay men living in China, Travis Kong (2011) reported that those who were born in the 1980s celebrated their differences and individuality. Comparatively, earlier cohorts had internalized homophobia and felt ashamed of their feelings. This shift partly occurred because younger generation growing up in the 2000s greatly benefited from the Chinese economic reform in three ways. First, economic reform brought in commercial media programs that publicly addressed the issue of homosexuality. Second, the establishment of consumption venues such as bars, karaoke lounges, and bathhouses offered physical venues where gay men could meet. Third, the rise of the internet facilitated gay men's exploration and development of their sexual identity. The earliest gay-oriented websites started appearing around 1998 (Ho, 2010). One of Kong's (2011) informants described the internet as "the real enlightenment" (p. 163) because it provided gay men with nonpathological information about being gay and suggested ways to connect to the larger queer community.

However, gay men accepting themselves and building their community did not translate into mainstream acceptance among the Chinese public. Even today, gay men are not completely free from social discrimination (Liu & Choi, 2006). A 2013 Pew Global Attitudes Survey found that only 21 percent of the population in China agreed that society should accept homosexuality, compared with 54 percent in Japan and 39 percent in South Korea (Pew Research Center, 2013).[6] Currently, social pressures primarily come from the workplace and family. In corporations and government departments, married men are considered to be more responsible and stable and are therefore more likely to be promoted. Gay men who are not married to women have fewer career advancement opportunities. Further, the Chinese family exerts tremendous pressure on marriage. Kong's (2011) informants had to devise various tactics to handle such pressure. Some avoided mentioning anything related to relationships or marriage. Others told their parents they did not have the time or money to marry. Still others engaged in "cooperative marriage," a fake marriage between a gay man and a lesbian that has become increasingly popular (Choi & Luo, 2016).

The advent of smartphones further shaped the way gay men connect with each other. Grindr, the gay dating app pioneer, was launched in 2009 in the United States, two years after the iPhone was released. In China, Blued, the first locally developed dating app for men who are interested in men, was founded in 2012 by Geng Le.[7] Following Blued, the local apps Zank and Aloha were launched in 2013 and 2014, respectively.[8] At the time of writing, Western apps such as Grindr and Jack'd can be used in China, but their connections are not stable; furthermore, because Grindr and Jack'd are available only in English, many gay men in China prefer using local apps that are in Chinese. Although nationally representative user statistics are unavailable, studies conducted by private analytic agencies and information released by the app companies indicate that local dating apps are extremely popular among gay men in China. In June 2015, Blued reached three million daily active users (Dou, 2015). One big data research agency found that in 2015, out of the pool of 120 million monthly active smartphone phone users it monitored, 460,000 used Blued, 207,000 used Zank, and 71,000 used Aloha (Analysys, 2016). All of these apps operate on geolocation information and provide "people nearby," "swiping," "status updates," and "live streaming" features (see chapter 1).

The overview provided above shows that, undoubtedly, the life of gay men in China has become easier in the last two decades due to decriminalization, depathologization, and economic reform. The increasing popularity of dating apps now offers gay men in China new ways to connect. Nonetheless, being gay is still not entirely socially acceptable in this country. Later in this chapter, I explain how users translate these complicated social circumstances into contradictory emotions through dating apps. But before I dive into the lived stories of my informants, I explain my understanding of affect and emotion to support my analysis.

AFFECTS AND QUEER POLITICS

Affect theory concerns the noncognitive component underlying embodied experiences. Rather than being a coherent theory, affect theory is a set of competing ideas without a clear consensus. Donovan Schaefer (2015) identifies two schools of affect theory—the Deleuzian school and the phenomenological school. The Deleuzian school, which includes theorists such as Gilles Deleuze, Brian Massumi, Patricia Clough, and Eric Shouse, contends that affect is pure intensity that structures our experiences. It is completely different from emotions in the sense that affect is a force our body feels before our consciousness recognizes it. To put it simply, once an affect can be named—that is, it is recognized consciously—it is no longer an affect; it becomes an emotion (Shouse, 2005).

The phenomenological school, which includes figures such as Eve Kosofsky Sedgwick and Sara Ahmed, uses the terms *affect* and *emotion* interchangeably. Their theoretical perspective emphasizes embodied experiences. The analytical distinction between affect and emotion, in Ahmed's words (2004a), "risks cutting emotions off from the lived experiences of being and having a body" (p. 39). Schaefer (2015) himself argues that the separation of affect and emotion in the Deleuzian school has created a dualism that Deleuzian philosophy is eager to push back against. More important, theorizing affect as pure intensity means reducing the phenomenologically diverse experience to an abstract, singular modality, similar to reducing different colors, which are phenomenologically unique to our senses, into pure light waves. My view of affect is more in line with the phenomenological school because the affects experienced by my informants in their use of dating apps were deeply social. Their emotions were heavily shaped by both the long social history of homosexuality in China and their immediate personal experiences with dating apps.

Influenced by the work of psychologist Silvan Tomkins, Sedgwick (2003) writes, "affects can be, and are, attached to things, people, ideas, sensations, relations, activities, ambitions, institutions, and any number of other things, including other affects" (p. 19).[9] Ahmed (2004b) further elaborates on how emotions can be "attached" to things. She refutes the idea that emotions come internally from the individual psyche or externally from things. Instead, she views emotions as cultural and social practices embedded in histories. According to her, they result from contacts between the individual and the social, which then "produce the very surfaces and boundaries that allow the individual and the social to be delineated as if they are objects" (p. 10). In this sense, emotions are performative because they generate new associations and maintain existing associations between the individual and the social. Following the logic of capital accumulation, Ahmed's "affective economy" describes how a certain emotion builds up through circulation. Emotions form through interactions between individuals and things, then further interactions are interpreted through these emotions.

Affect theory is intimately related to queer studies through feelings of shame and pride. A profound articulation of shame and queer lives was put forth in Sedgwick's influential essay "Queer Performativity: Henry James's *The Art of the Novel*" (1993). In it she writes, "if queer is a politically potent term, which it is, that's because, far from being capable of being detached from the childhood scene of shame, it cleaves to that scene as a near-inexhaustible source of transformational energy" (p. 4). Citing Tomkins, Sedgwick argues that shame delineates identity because "shame, as opposed to guilt, is a bad feeling that does not attach to what one does, but to what one is" (p. 12). Jon Elster (1999) defines shame as "a negative emotion triggered by a belief about one's *own character*" and guilt as "a negative emotion triggered by a belief about one's *own action*" (p. 21, emphasis added). Queer shame is based on a convincing belief that the heteronormative world is right and that being queer is undeniably wrong. To live a queer life, therefore, is to perpetually fail to meet the norms of the heteronormative world. Comparing the regulative norms of heteronormativity to "repetitive strain injuries," Ahmed (2004b) contends that "through repeating some gestures and not others, or through being oriented in some directions and not others, bodies become contorted" (p. 145). Queer bodies hurt when they are forced to conform to heteronormative molds.

Since the 1969 Stonewall riots in New York, pride has become a major rhetoric used by gay rights activists to counter the mainstream discourse where gay men are ridiculed for their abnormal sexual practices and blamed for causing the HIV/AIDS outbreak. For some activists, pride has been treated as an antidote to shame. Elster (1999) defines pridefulness as "a positive emotion triggered by a belief about one's *own character*," whereas pride is "a positive emotion triggered by a belief about one's *own action*" (p. 22, emphasis added).[10] Similar to Ahmed, Erin Rand (2012) suggests that emotions are social and historical products, not simple psychological traits. The emergence of gay pride in the United States was tightly connected to sociopolitical circumstances. In the United States, citizens are protected by the First Amendment, which grants them freedom of speech and the right to peaceful assembly. However, the assimilationist stance and neoliberalization of gay pride have frustrated some American queer activists and scholars, who from the late 1990s started organizing "Gay Shame" to revive shame as a ground for collective resistance (Halperin & Traub, 2009).

Whether the discomfort felt by queer bodies about shame or pride in the gay movement, affects have often been ignored in traditional disciplines that have focused on social structures, economic disparities, and institutional arrangements. Heather Love, in an interview with Sarah Chinn (2012), said, "Without attention to affect I think it's a real struggle to articulate and explain the way that oppression registers at small scales—in everyday interactions, in gesture, tone of voice, etc." (p. 126). Therefore, in the following analysis, I take up this challenge by discussing emotions that my queer male informants articulated when they described their experiences with dating apps. I am not concerned with whether the emotions I identify are basic or synthetic emotions or whether I have produced an exhaustive list. After all, even psychologists of emotion disagree on how many basic emotions there are.[11] What is important is not the exact number of basic emotions we have but what emotions do (Ahmed, 2004b). Some emotions pull people closer to an object, while others push them away.

From the rich personal narratives of my queer male informants, I differentiate in-app emotions, which are directly derived from the everyday use of dating apps, from out-of-app emotions, which are rooted primarily in the way male homosexuality is treated in contemporary Chinese society. I also identify positive emotions that drive users closer to dating apps and negative emotions that pushed them away from the apps. Dichotomizing emotions into positive and negative may appear to oversimplify the complexity of human emotions. However, I argue that positive and negative emotions are not mutually exclusive because my informants oftentimes experienced several emotions within a very short period of time. Identifying the valence of emotions also delivered analytical value. Understanding these contradictory emotions and the oscillation of positive and negative feelings is vital to the puzzle I outlined at the start of this chapter: what led my queer male informants to cyclically install and delete these dating apps?

IN-APP EMOTIONS

POSITIVE EMOTION: JOY

During our interviews, my informants often expressed moments of joy when they recalled their experiences using dating apps. With an app's global positioning system, users can find other queer men who are physically nearby. Roderic Crooks (2013) calls the virtual space created by these apps an "ad hoc social space." Yoel Roth (2014), in his critical reading of Scruff, a popular app in Western countries, suggests that this overlaying of a virtual, digital space on top of an actual, physical place created a heterogeneous space where queerness can coexist with heterosexuality. A McDonald's, a church, or a cafeteria can immediately be turned into a gay space with these apps.

My informants generally agreed that dating apps had allowed them to connect with each other. Gay bars in Guangzhou are scattered around the city and are often found in hidden locations. Many of my younger informants either did not know where these bars were, or if they had visited them, they preferred not to speak with strangers. In addition, consumption at these venues was not cheap. A standard cocktail in a well-known gay bar or restaurant in the city cost CNY60 (~USD9), which was considered expensive because many of my informants made CNY5,000 (~USD725) or less per month. Therefore, dating apps provided them with an economical way to meet other queer men in the city.

Gay dating apps also tell their users the exact physical distance between themselves and others (figure 4.1).[12] This indicates the likelihood of an offline interaction because greater distances reduce the intention to meet. Commenting on the dating app Blued, the gay-identifying man named Green, age twenty-four, said, "The app can easily locate people who are nearby, making it easier to meet face to face for chit-chat or a coffee. So I think it can help me find a person whom I can have a great conversation with." Similar to my heterosexual informants, distance matters more in romantic relationships. Yuan, age twenty-six and gay-identifying, commented, "Distance matters. I think five kilometers are still acceptable, but some people think it is already too far. . . . But beyond 10 kilometers is like, a long-distance relationship."

Although dating apps afford visibility and authenticity, users can choose to remain anonymous. On registering for a dating app account, users are asked to set up their profiles by uploading photographs and filling out their personal information. Some Western apps, such as Tinder, link users' profiles to their Facebook accounts. However, most gay dating apps in China do not require users to disclose their information

Figure 4.1
Blued shows the exact distance between users. (Screenshot taken by the author on July 21, 2018)

on their profile. For example, Blued allows users to upload photographs and asks for ten pieces of demographic information; however, users can either leave this blank or provide fake information (figure 4.2). Blued users can apply for an official verification of their face photograph. To accomplish this, they must upload a video that clearly shows their face on the platform. The platform then compares this video with their other photographs on the app. Once verified, a check mark will appear next to the profile. However, this feature was not popular among Blued users when I conducted my research.[13]

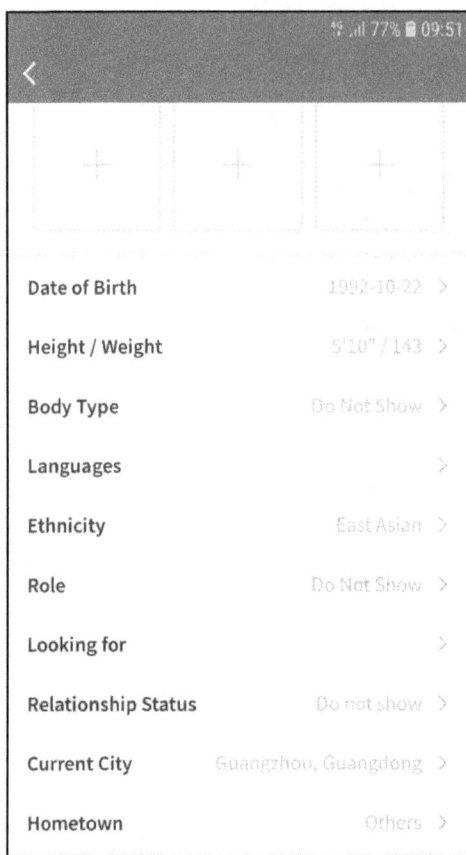

Figure 4.2
Blued allows users to upload photographs and asks for ten pieces of demographic information. (Screenshot taken by the author on July 21, 2018)

Given the ability to hide their identity, my queer male informants said they felt freer to exercise their sexual desires. For instance, Yuan had experienced cybersex via his smartphone's camera with people he had met on Blued. He accepted this practice only because "this won't leak information about your everyday life." Even with people he had known for some time, he still hid his face from the video, pointing out that he was "wearing a mask to act."

Compared with previous decades, the modern era and its technological progress had given my informants supreme anonymity. In earlier eras, queer men relied on subtle body language and accessories (such as earrings or handkerchiefs) to determine the sexual orientation of someone they encountered in a public space, and police raids of public bathrooms were commonplace. With dating apps in their hands, queer men did not have to worry about misrecognizing a straight man as gay or being caught by the police when they sought casual sex. These rewards sustained my informants' continued use of the dating apps.

NEGATIVE EMOTIONS: ANGER AND DISGUST

In-app experiences are not always positive. A common complaint among my informants was that using dating apps risked being deceived because the users of online dating sites often misrepresented themselves (Ellison, Hancock, & Toma, 2012; Whitty, 2008). In Chinese dating app culture, one's *yanzhi* 颜值 (literally, the "the value of one's face") (S. Wang, 2020) has a significant impact on how popular one is on an app. Therefore, the use of fake photographs or photograph enhancement is commonplace. Xiaoming, age twenty-one and identifying as a male *tongzhi*, once encountered a man impersonating someone else: "I chatted with person B before and person B sent me a photo of his. Then person B talked to person A and sent his photo to person A as well. When I chatted with person A, he sent me person B's photo." Xiaoming was outraged at this transgression. Experiences like this made my informants extremely suspicious of others' photographs. For instance, Erza, age twenty-four and identifying as a man who desires other men, estimated that out of ten profiles on Blued, three or four used fake photographs and another three or four used edited photographs. Prior research has found that if the discrepancy between the online persona and the offline persona falls within a reasonable range, it is often tolerated or even expected (Ellison, Hancock, & Toma, 2012). However, my informants considered fake photographs unreasonable and were greatly irritated by them.

Back in 2014, when I started conducting research on dating apps in China, I encountered an interesting profile on Jack'd—a complaint.[14] The profile creator used photographs of Ye Haiyan, a prominent Chinese feminist activist. Under the "About me" section, this creator criticized another's online profile for providing fake information:

There is a person sending out fake photos every day. He keeps changing his information, sometimes in Shenzhen, sometimes in Wuhan, and
sometimes in Hong Kong. His age, height, and weight are changed all of
the time. The only thing that he does not change is asking for dick pics
from others. Don't you feel tired? . . . Can't you be sincerer? You are not
handsome; you are not outstanding. No one is interested to know who
you really are. If you really regard yourself as a big gun and are afraid of
fans' following you, you don't have to send photos. Sending fake photos
and, what's worse, web photos that others have been sending out for some
time? Don't you think you are very "low"?[15]

This profile highlights the prevalence of fake information on dating
apps and indicates how angry the writer of the critique was. Further, it
speaks to another issue that my informants found disgusting—harassment.
In chapter 2, I pointed out that my straight female informants received
unsolicited sexual requests from men on dating apps. My queer male
informants shared similar experiences. In the words of Huajun, who was
twenty-eight and identified as gay, on dating apps "you cannot choose to
be disturbed or not to be disturbed."

Harassment occurred pretty often on Blued because the app allows
users to send messages to anyone at any time. Allen, age twenty-four and
identifying as gay, lamented,

I met a forty-year-old "uncle."[16] He was extremely disturbing. As long as
you were on Blued, you would notice that he sent you a message every two
days. You had already clearly rejected him, but he still kept messaging. . . .
He also sent photos. . . . I told him not to disturb me anymore, but he did
not listen. Eventually, I blacklisted him.

Harassing messages can be very direct. The most common one my
informants received was yue? 约? (meet?). This one-word message
implied meeting for sex. Messages can be vulgar, too, with language like
"I want to suck your dick" or including an unsolicited "dick pic." Some
of my informants even encountered harassers who teased them for being
naïve because they were looking for friends on these apps. In their critical
analysis of the prevalence of dick pics in Western digital culture, Susanna
Paasonen, Ben Light, and Kylie Jarrett (2019) write, "dick pics are pervasive within dating and hook up apps used by same-sex attracted men and
are a generally accepted actor within this sexual infrastructure" (p. 6). Yet
dick pics are definitely not common in Chinese gay dating app culture,

at least from my informants' perspective. The ubiquity and acceptance of dick pics to which Paasonen et al. refer do not match up with the experiences in China, and such overgeneralizations may lack cultural specificity.

Another issue pertaining to geoinformation is that locative information is available to *both* parties on dating apps. This feature brought another layer of discomfort to my informants. Damon, age twenty-five and identifying as gay, explained to me that when someone knew how close he was, he had difficulty rejecting an invitation. "He and I live in the same building, and he has constantly invited me to hang out or visit his home. It is so difficult to reject." Eventually, Damon stopped using Blued and switched to an app that showed him only people from different cities.

My informants regarded such harassing and annoying behaviors as "low" or lacking *suzhi* 素质 (quality), as indicated in the profile I translated above. The word *suzhi* "justifies social and political hierarchies of all sorts, with those of 'high' quality gaining more income, power and status than the 'low'" (Kipnis, 2006, p. 295). In the Chinese gay culture, the *suzhi* discourse has often been used to discriminate against so-called money boys who come to the city from rural China (Ho, 2010; Kong, 2011; Rofel, 2007). My informants, however, used this term less to refer to money boys than to describe older gay men who bluntly, explicitly, and persistently solicited sex from them even though they had already clearly rejected such solicitations. My informants did not object to seeking casual sex on dating apps or being propositioned. Rather, they objected to being disturbed again after they had rejected the initial solicitation. The lack of *suzhi*, in this context, refers to an online etiquette that the older app users failed to comply with. The case of Allen and his forty-year-old admirer illustrates the point.

OUT-OF-APP EMOTIONS

POSITIVE EMOTION: HOPE

Apart from joy, my informants' experiences revealed an additional layer of emotion emanating from their experiences of living as gay men in contemporary China. This was the hope of gaining social acceptance.

As I mentioned above, the dominant rhetoric in Western LGBTQ activism is pride. However, due to limited freedom of speech and assembly in China, large-scale LGBTQ events are rarely seen. For example, the Beijing Queer Film Festival, which started in 2001 as the Beijing Gay and Lesbian Film Festival, was repeatedly raided by the police until it was eventually stopped in 2014 (P. Fan, 2015). The award-winning gay romance *Call Me by Your Name* was suddenly removed from the Beijing International Film Festival by the organizer in 2018. During my fieldwork, Guangzhou Gay Pride events were canceled just a few days before they opened. One of my informants told me that the event could proceed only as an underground private affair. In 2016, China voted in the United Nations (UN) Human Rights Council against the establishment of the mandate of the UN Independent Expert on protection against violence and discrimination based on sexual orientation and gender identity. In 2019, it again refused to recognize rights by voting against the mandate's renewal. Thus, if feelings of pride are a reaction to the things that someone has done (Elster, 1999), currently the local Chinese queer community has little to feel proud of. It has made no significant accomplishments and failed to enact legal protections from discrimination.

However, hope works on a different logic. It looks to the future, not the present. Simply put, hope is the expectation that something good may happen (Elster, 1999). Precisely because it is based on something that has not yet been realized, hope is difficult to completely disregard. For this reason, it is a highly resilient emotion. In *Cruising Utopia: The Then and There of Queer Futurity*, Jose Muñoz (2009) notes that even though negative emotions can bind people together, hope is what drives LGBTQ people to pursue utopia in the midst of their current unpleasant situations. Although there is no doubt that my queer male informants felt tremendous social pressure, there were sporadic events that gave them hope. For example, in April 2018, Weibo, a major Chinese Twitter-like platform, announced that it would ban homosexual content in a supposed effort to clean up the online environment. Such a ban touched the nerves of both the LGBTQ community and its advocates, who immediately criticized this policy through hashtags. Four days later, the Weibo platform reversed the ban. Even *People's Daily*, the official newspaper of the CPC, responded to the ban by encouraging tolerance toward queer people ("Online Outcry," 2018). In Shanghai, where the LGBTQ culture is striving, Shanghai Pride and three separate queer film festivals (Cinemq, ShanghaiPRIDE Film Festival, and Shanghai Queer Film Festival) have been held in the city every year. Although these events are all run by volunteers and are "tolerated rather than supported by the local government" (Newby, 2018, para. 6, describing Shanghai Pride), they have helped to build a more tolerant city by facilitating conversations between the general public and the LGTBQ community.[17]

My informants regarded the existence of gay dating apps as an indicator that their community would eventually be accepted. Jerry, age twenty-eight and identifying as queer, said, "Although these apps are not a mainstream thing, their existence and the online interactions they support allow Guangzhou *tongzhi* to survive. . . . [Dating apps] represent our existence. . . . Apps give me hope." The hope that Jerry articulated did not originate with his use of the dating apps. Rather, his strong affective response came from the symbolic meaning he associated with these apps. After living in New York for seven years, he believed one day queer people in China could freely live their queer lives in a way similar to their American counterparts. Dating apps represented the potential for a queer community that he could identify with. Johnny, age twenty-seven and identifying as a male *tongzhi*, said that the existence of dating apps meant that queer men no longer needed to "live underground." To him, dating apps provided the queer community with a certain level of legitimacy that one day would blossom into societywide acceptance. River even compared gay dating apps to the Christian Gospel. In the Gospel, faith in Jesus would bring believers eternal life in the future. To River, gay dating apps would bring him a gay community in the future. Although many of my queer male informants, including Jerry and River, were affiliated with LGBTQ-serving organizations in Guangzhou, they recognized that dating apps were an indispensable platform for developing a queer community.[18]

NEGATIVE EMOTION: FEAR

Although anger and disgust were negative in-app emotions caused by the prevalence of deception and unsolicited requests for sex, the most dominant negative emotion my informants expressed in relation to dating app use was rooted in the social stigma of male homosexuality in China. They feared their family members and colleagues would discover they were gay. The opposite of hope, fear comes from expecting bad things to happen in the future (Elster, 1999). The geoinformation feature on dating apps, while allowing users to locate each other, also brings the risk of being involuntarily outed. In the early 1990s, being outed as gay was the worst possible nightmare for the professional future of American politicians (Gross, 1993). Likewise, in China today, being outed can have an extremely negative impact on a person's career and family relationships, even though homosexuality is no longer considered a crime or a mental disorder.

Many of my queer male informants had not come out to their colleagues. Among the queer-themed public accounts I followed on WeChat since the beginning of my fieldwork, dozens of people had explained that for career advancement, marrying and then divorcing was better than remaining single, particularly for those who worked in government or state-owned enterprises. My informants who worked in traditionally "masculine" occupations particularly expressed concern over being outed. Tesla, age twenty-four and gay-identifying, worked as a software engineer and worried about workplace discrimination. Huajun, who was a personal trainer, often had to blend in with his straight male colleagues by making sexist jokes. In all cases, it was easier for these queer Chinese men to align their public appearance with hegemonic masculinity than risk the professional and personal consequences.

Similarly, most of my informants did not plan to come out to their parents. In traditional Chinese culture, a son's most important duty is to continue the family line (Fei, 1939). Mencius once said, "There are three ways of being unfilial, and to have no posterity is the greatest of them."[19] Because these historical expectations loom so large, Chinese parents who learn that their son is gay are often devastated. Lasong, age twenty-six and identifying as a male *tongzhi*, told me that he did not upload any face photographs to dating apps. He did not want his cousin, who he knew was also gay and lived near him, to identify him. Instead of thinking that he and his cousin would become closer if they mutually recognized each other's sexual orientation, he was afraid his cousin would "betray" him and tell his family. Jerry was one of the few queer men I interviewed who had come out to his mother. On hearing his revelation, his mother went to multiple hospitals to find out whether she could conceive a second child, which demonstrates how critical continuing the family lineage was to her.

Like hope, fear is resilient because it is an affective response to unpleasant events that have not yet happened. None of my informants had been involuntarily outed because of their use of gay dating apps. However, as long as they believed there was a chance that they would be exposed as gay, their fear remained. Shame, once a major rhetoric in Western gay history, did not surface in my interviews. I believe that this was because my queer male informants had come of age after the decriminalization and depathologization of male homosexuality in China. None of them saw themselves as immoral or wrong. And although they continued to fear being outed, they also saw their queer identities as a source of power to eventually change the minds of friends, family, and colleagues. They acknowledged that their families or workplaces might not endorse their sexual orientation, but they regarded this lack of endorsement as social conservatism they could push back on.

AFFECTIVE FABRICS OF DATING APPS AND QUEER POLITICS

The analysis of my informants' experiences revealed a contradictory set of emotions in relation to dating apps. Discussing digital culture, Adi Kuntsman (2012) proposes the concept of "affective fabrics" to describe "the lived and deeply felt everyday sociality of connections, ruptures, emotions, words, politics and sensory energies" (p. 3). Kuntsman urges us to pay attention to how "affect and emotions take shape through movement between contexts, websites, forums, blogs, comments, and computer screens" (p. 1). Although my analysis focused on only one type of digital platform, it was clear that the contradictory emotions my informants felt were not only derived from their everyday use of dating apps but specifically reflected out-of-app queer politics in China. Together, these emotions formed an affective fabric that both held them back and propelled them forward.

The affective fabric of dating apps for my queer male informants was interwoven with both smooth and coarse emotional threads. The smooth part of the fabric was made with joy and hope. Locating other queer men who were otherwise invisible was once an impossible task for Chinese urbanites. Dating apps undoubtedly facilitated queer men's networking. Users can now anonymously look for casual relationships or set up a full profile to look for serious relationships. As for hope, as I mentioned, in Guangzhou there is no gay neighborhood where queer men can hang out safely and freely. Therefore, dating apps offer a possibility for a "queer village" (Crooks, 2013).

The coarse thread was primarily made up of anger, disgust, and fear. Although I have empathy for users who fabricate their identities on the apps to protect themselves, no one wants to be deceived. The threat of being deceived may perpetuate a vicious cycle of mistrust. When users experience deception or hear stories about their friends being deceived, they may lose trust in others. If they think people on an app are dishonest, why would they put their authentic information and photographs there? The distrustful environment and deceptive behaviors take on a life of their own in the behaviors of queer men who decide to remain closeted. Further, Chinese dating app culture is still nascent. Norms for interactions, including when it is appropriate to send a sexually suggestive photograph, have not yet been fully established (R. H. Jones, 2005). Unsolicited and incessant harassing messages induce disgust among some users. The fundamental fear related to dating apps is the risk of being outed. Until the day that society fully accepts homosexuality, gay dating app users will continue to be worried about being involuntarily exposed to family and colleagues.

Although positive emotions drove my informants closer to a dating app, negative emotions drove them away. In this affective economy, every new experience with a gay app charged the app user with stronger contradictory emotions. Slowly, the app became what Ahmed (2010) refers to as a "feeling-cause"—an object that can cause feeling. In this case, the feeling was ambivalence. One way my informants managed this feeling was by constantly installing and uninstalling the same app. This sense of ambivalence was also apparent from the long silence that usually followed when I asked my informants to evaluate the influence of dating apps in their life. It was not easy for them to provide a conclusive evaluation because they were grappling with contradictory emotions. Comparing the social environment of queer people with that of their straight counterparts, the latter did not have the same ambivalent feelings toward the apps they used. I am not arguing that there were no contradictory emotions induced by their use of dating apps. From my earlier analysis, it is obvious that my straight female informants had both positive emotions when exploring their sexuality and the world and negative emotions when facing sexual harassment and surveillance. The straight male informants were annoyed by bar promoters and impersonators and simultaneously enjoyed presenting themselves in a better light to maximize their opportunities for romance and casual sex. But these emotions were, inescapably, all in-app emotions. Out-of-app emotions, which are tightly connected to queer politics, were not relevant to them at all. In our heteronormative society, my straight informants had no fear of the stigma of being gay. Neither did they need to hope for social acceptance. Both fear and hope—the more resilient emotions—were experienced only by my queer informants. Therefore, it is little wonder that the feeling of ambivalence was more dominant among my queer informants than among my straight informants.

CONCLUSION

This chapter begins with a question: why did my queer male informants keep installing and uninstalling the dating apps they were using? Existing scholarship tackled this question primarily by appealing to users' cognitive evaluation of the technology's functions. My approach, by comparison, paid attention to my informants' affects and emotions in relation to dating apps. Relying on the phenomenological school of affect theory, I have shown that these emotions were tightly related to how homosexuality is viewed in China today. Using dating apps triggered multiple contradictory emotions among my queer informants. On the one hand, it allowed them to conveniently locate other gay men and to safely exercise their sexual desires. They saw these apps as a symbol of the future gay community. To them, these apps were an object of joy and hope. On the other hand, dating app use induced the fear of being outed, anger at being deceived, and disgust over sexual harassment. Dating apps simultaneously carried negative emotional charges. By addressing this affective aspect of dating app use, my discussion moved beyond attributing the installing or uninstalling of apps to sheer usefulness or ease of use. My analysis showed that there were deep social and historical roots underpinning the use and nonuse of dating apps among gay Chinese men, a circumstance my straight informants did not share. My queer male informants' feelings toward dating apps, therefore, are most accurately described as a manifestation of contemporary Chinese queer politics. The dominance of fear and hope in relation to dating apps, rather than shame and pride, reflects the very different trajectories of queer activism and rights in China and the United States.

In the next chapter, I shift my attention to my queer female informants, my final set of networked sexual publics. Because of their unique social positions, they collectively told me a very different story. Compared with my queer male informants, they did not express significant ambivalence toward the dating apps. Instead, many of them held these apps in high regard. They thought of dating apps as contributing to their community. I explore the reasons behind this difference. Is it because the features of lesbian apps are different from gay apps? Is it because queer men and women have different subcultures? Or are queer women and queer men simply caught in different binds?

5

BUILDING A CIRCLE FOR QUEER WOMEN: AFFORDANCE OF COMMUNAL CONNECTIVITY

I always asked my informants how they would feel if the dating apps they had been using were to disappear the next day. This question was not entirely hypothetical. Zank, once a popular app for queer men, was permanently shut down in 2017 by the Office of the Central Cyberspace Affairs Commission due to its pornographic content. Rela, an app for queer women, was temporarily taken off the shelf the same year, allegedly because of its involvement in an organized protest for marriage equality in Shanghai.[1] Shawn, who was twenty-four and identified as queer,[2] responded to my question as follows:

> I think [dating apps] more or less represent the legitimacy and legality of our lesbian *quanzi*. For me, their value is greater than their use value. I am very glad that they exist, and I cannot accept the idea of their being taken off the shelves.

Similar to Johnny in the last chapter, Shawn saw lesbian dating apps as a symbol of legitimacy for same-sex intimacy. Xiaoyan, age twenty-eight and identifying as lesbian, also said that the disappearance of lesbian dating apps would be very upsetting:

> Apps like Rela and Lesdo are built for the lesbian *quanzi*. In China, the lesbian population is small. When you develop an app, you want to make money, right? But even although you know the market is small, you still develop the app. It is probably because you are passionate and altruistic.

These apps are well designed, and I cannot deny that because of these apps, people in the lesbian *quanzi* know that they are not alone in this world. . . . Therefore, if these apps disappeared, I would be very sad.

Both Shawn and Xiaoyan mentioned the term *quanzi* 圈子 (circle). The term is commonly used by Chinese LGBTQ people to represent the group they belong to. It can refer to a generic community or "an immediate small network of close friends, a sense of shared socioeconomic location or way of life for a wide demography" (Engebretsen, 2014, p. 133). Its contextualized and relative nature is similar to the idea of "publics" articulated by Sonia Livingstone (2005; refer to chapter 1). Both *quanzi* and publics entail a form of emotional connection with a collective that is beyond a dyadic relationship. In talking with my queer female informants, I found this *quanzi* aspect extremely salient in their use of dating apps. I am not suggesting that my queer male informants did not speak about their *quanzi* in their narratives. In fact, the discussion of hope in the last chapter highlights the potential that queer men saw in gay dating apps to bring forth a visible, legitimate gay community in China. However, my queer male informants tended to emphasize dyadic—sexual or romantic—connections. My queer female informants, as I show in this chapter, reported using dating apps such as Rela and Lesdo not only for romance and casual sex but equally for emotional support, information, and advocacy.

Recent research on queer digital culture has explored the role of social media in queer world-making (e.g., Cavalcante, 2019; Hanckel, Vivienne, Byron, Robards, & Churchill, 2019). Researchers have asked if dating apps have helped to build the queer communities, with mixed results. For instance, Sam Miles's (2017) in-depth interviews with queer men living in London show that only heavy app users perceived dating apps to be useful in reconstituting the fragmented gay communities in the city. Light users said they did not share anything in common with the other users except that they used the same app. In her study of queer women's use of Tinder in Canada and Australia, Stephanie Duguay (2019) argues that Tinder fails to foster a sense of queer community. She traces this shortcoming to a lack of affordances that enable interactions between queer women and little interest among queer women in meeting women who were physically far away. Conversely, the informants in

Thomas Baudinette's (2019) ethnographic study of Tokyo's gay culture held dating apps in high regard. Dating apps drew them into Ni-chōme, a well-established gay district, and reinforced the "gay feeling" of that neighborhood. His informants said that dating apps were so critical to this "gay feeling" that "a visit to Ni-chōme would be incomplete without using dating applications" (p. 97).

Following this line of inquiry, this chapter elaborates on how dating apps such as Rela and Lesdo have enabled my queer female informants to connect with and constitute their *quanzi*. My focus on the social aspects of my queer female informants' use of dating apps should not be read as perpetuating the stereotype that queer women are not interested in sex, as apps can fulfill both sexual and social needs. My decision to narrow down to the social aspects of queer women's app use was based on the fact that, compared to other networked sexual publics, the issue of *quanzi* stood out in the accounts of my queer female informants. As I show later, the underemphasis of hookup experiences in their accounts is likely to be a social product of both the apps' features and the patriarchal scrutiny they are living in.

This chapter's focus on queer women was further informed by intersectional thinking (Crenshaw, 1989; Hancock, 2016). Although dating app studies are growing, we have not yet adequately addressed the unique experiences of queer Chinese women. Only a handful of studies have focused on queer women's experiences with dating apps (e.g., Choy, 2018; Duguay, 2019; D. T.-S. Tang, 2017). Further, when queer women have been recruited as research participants, they have often been lumped together with queer men or the general population. Kimberlé Crenshaw's (1989) work on "intersectionality" describes how race and gender intersect to produce social injustice against African American women. Likewise, intersectionality-like thinking helps me to highlight the unique struggles queer women face as they occupy the intersection of two subordinate positions.

In the following, I first describe the distinct set of challenges faced by queer women living in China. As I mentioned in the introductory chapter, the concept of "affordance" enables a more nuanced understanding of the relationship between communication technology and society (Baym, 2010; Hutchby, 2001; Neff, Jordan, McVeigh-Schultz, & Gillespie, 2012).

Based on my analysis of the design and users' experiences with the two lesbian dating apps, I explore a salient affordance of these apps: communal connectivity. I argue that dating apps have enabled queer women to connect with their *quanzi* even as they also reproduce heteronormativity.

THE DOUBLE PREDICAMENT FOR QUEER WOMEN

In the historical research on Chinese female same-sex desire, there have been two ongoing interrelated debates. The first discussion is how prevalent female same-sex relationships were in premodern China, while the second involves whether these relationships were tolerated. In one of the earliest historical studies on Chinese sexuality written in English, Robert van Gulik (1961) suggests that female same-sex relationships in ancient China were not only prevalent but also tolerated. Based on Daoist sexual manuals and historical literature canons, van Gulik contended that female same-sex relationships were commonly found between wives or concubines in polygamous marriages. Historical descriptions of the lives of women in the palaces also show the prevalence of *duishi* 对食 (paired eating), a practice where two women live like a husband and wife (Hinsch, 1990; Shi, 2014). Van Gulik's claim that female same-sex relationships were common, however, is objected to by historian Bret Hinsch. In his work—which primarily explores the history of male homosexuality— Hinsch (1990) points out that compared with references to male homosexuality in historical sources, references to female homosexuality were rarer. He writes:

> Partly this lack was due to the relative absence of personal freedom accorded women. Bound to their husbands economically and often forced into seclusion in the home, many women were denied the opportunities to form close bonds with women outside their household. (p. 173)

The disagreement between van Gulik and Hinsch can be reconciled by considering the baseline for comparison. Hinsch's point of comparison was the number of references to male homosexuality in historical records, which led to his conclusion that female homosexuality was less common than male homosexuality. Also, the different corpuses that these authors used in their analyses affected how frequently female same-sex relationships would have been referenced. References to female same-sex relationships have often appeared in the minor literature, instead of the canons that van Gulik and Hinsch relied on. For example, based on her analysis of such literature from late sixteenth- to twentieth-century China, Tze-Lan Sang (2003) argues that there has been more material on female same-sex relationships than we might have thought. For instance, husband's wives having sex with each other was a common theme she found in erotic literature written by male authors.

Can this recurrent motif be an indicator of female homosexuality being tolerated in premodern China? This brings up the second issue on which scholars of Chinese sexuality have disagreed. When van Gulik found references to female same-sex practices in Daoist sexual manuals, he assumed this meant female same-sex practices were accepted. Wah-shan Chou (2000) argues that because Chinese traditional culture has never explicitly rejected homosexuality, there is a cultural tolerance of it, including female homosexuality. Sang (2003), however, fundamentally rejects both claims. The erotic literature she included in her analysis always situates female same-sex relationships within a heterosexual marriage. The typical story in this literature is that two wives love each other, so they are not jealous of each other even if their husband favors one over the other. However, when two women love each other to the extent that they cannot accept a heterosexual marriage, they generally face death. In this sense, polygamous marriage is portrayed as a way for women to cover up their same-sex desires in this literature (Sang, 2003), which reflects what Adrienne Rich (1993) calls "compulsory heterosexuality." Liang Shi (2014) writes, "the fact that women find it necessary to hide female-female love under the cover of heterosexual marriage points to a milieu that forbids the expression and existence of such love" (p. 57).

In the 1920s, the Western discourse of sexual liberation and the idea of homosexuality were introduced to China (see chapter 4 for more details). Although Chinese intellectuals began to acknowledge women's right to derive pleasure from sex with their male partners, they continued to assert that it was unacceptable for women to receive pleasure from same-sex behaviors because they might resist marriage (Sang, 2003). Shi (2014) locates several fictional accounts from this era where lesbians are denigrated and dismissed. For example, there is a story about a female-serving female prostitute. In this story, the prostitute is referred to as "one of the eight anomalies of Shanghai," who tricks her clients into giving her all their money. Other narratives call queer women's communities *mojing-dang* 磨镜党 (mirror-rubbing gang), suggesting these women are violent and ganglike. Historically, where lesbians were acknowledged, they were often depicted as predatory villains.

These two debates—how common female same-sex relationships have been and whether they were tolerated in premodern China—remain controversial because the limited historical references resist a definitive conclusion. However, scholars have universally agreed that Chinese queer women have been less visible than their male counterparts in representations of all kinds and in real life. In news coverage of LGBTQ issues, Chinese journalists have often focused on the gay subculture (Sang, 2003). Lesbians have rarely appeared in independent movies.[3] Shi (2014) can identify only four local productions since 2000 in which female same-sex relationships appear as a major plot. He also observes that even in the Beijing Queer Film Festival (formerly known as the Beijing Gay and Lesbian Film Festival), movies about gay men outnumbered those about lesbians by a ratio of more than four to one.

In terms of real-life visibility, although gay public culture in China has a substantial spatial manifestation (for example, in parks, bars, bathhouses, and public bathrooms), queer women have a less spatially dominant public culture (Shi, 2014; Liu & Lu, 2005). The "three obediences" (*sancong* 三从) in Confucianism, which I discuss in chapter 2, required a woman to obey her father, her husband, and her sons. Elisabeth Engebretsen (2014) argues that these precepts have historically discouraged women from developing an independent presence in public spaces. Further, lesbian activism has often relied on the politics of community, not on the politics of visibility, which has prioritized its integration into mainstream society by "de-emphasizing difference and highlighting commonality and similarity" (p. 126). Lesbian rights organizations have also lacked financial resources. As men who have sex with men are regarded as being high risk for HIV in China ("2018 nian Zhongguo aizibing ganran renshu," 2018), local gay rights organizations can frame themselves as HIV-prevention advocacy groups, which helps them apply for funding from the state. Lesbian activism lacks a solid link with the HIV-prevention effort and therefore has been unable to secure stable financial support from the state (Engebretsen, 2014).[4]

Apart from public culture, Lucettta Kam (2013) elaborates on the prevalence of queer women marrying men. Chinese queer women face the same parental pressure to marry as queer men. However, returning to intersectionality's need for specificity when evaluating oppression, women are placed in a different situation. For a queer woman, marriage is the most legitimate way to gain autonomy from her parents. At the same time, it puts her on an undesirable life trajectory. Therefore, some of Kam's informants lived a secret dual life, marrying a man while maintaining an extramarital relationship with a female partner. Those who were more fortunate had husbands who agreed to an open relationship. Others went further, entering into a "cooperative marriage" with a gay man. Kam views the latter as a tactic lesbians used to demonstrate their "public correctness," a set of socially correct behaviors queer women must perform to secure public recognition.

In summary, although female same-sex relationships have existed in China since ancient times, they have been confined to either a polygamous marriage in the past or a monogamous marriage today. When female homosexuality was not subsumed under a heterosexual relationship, it received heavy criticism. By virtue of being women, lesbians are also restricted in their individual ability to create a visible public culture in everyday life and have limited collective capacity to seek financial support from the state. Female homosexuality has never been criminalized in modern China to the same degree as male homosexuality. Still, sitting at the intersection of two oppressed identities—female and queer—queer women encounter more challenges than straight women and queer men combined. With this double predicament, how can popular lesbian dating apps enable greater agency? How do these apps contribute to the making of their *quanzi*?

COMMUNAL CONNECTIVITY OF LESBIAN DATING APPS

In the introductory chapter, I trace the origin of the concept of affordance. Its transitions across multiple academic fields have generated fruitful discussions and applications of the concept but have also created conceptual ambiguities (Evans, Pearce, Vitak, & Treem, 2017; Gibson, 1979; Hutchby, 2001; Nagy & Neff, 2015; Norman, 1988). Yet scholars generally agree that an affordance is not a feature but is something that results from the interactive relationship between the features and the users of an object. That is, the constitution of an affordance concomitantly depends on the materiality of the technology and the users' interpretation. Due to this contingency, the outcome of an affordance cannot be predetermined. An affordance is, as Andrea Scarantino (2003) puts it, the "promise" that an object offers to us. A promise is a potential effect that is yet to be actualized, thereby highlighting the nondeterministic role of technology in our behaviors and society.

In the introductory chapter, I also review the five affordances of dating apps, which I identified in my earlier work (L. S. Chan, 2017b) as mobility, proximity, immediacy, authenticity, and visibility. When I first theorized these affordances, I was interested in why straight men and women in the United States used dating apps for dyadic relationships. I mainly was thinking about Western apps such as Tinder and Coffee-MeetsBagel. Therefore, these affordances were formulated in relation to building dyadic relationships, whether sexual or romantic, long-term or short-term, involved or casual. Because affordances depend on both the features and the users, my research on Chinese lesbians and their dating apps revealed a different kind of affordance that I had not considered earlier. I call this affordance *communal connectivity*. In the following, I analyze the features of the two most popular apps for queer women in China, Rela and Lesdo, together with the interpretation of these apps by my queer female informants and the actual uses of these apps.

FEATURES OF RELA AND LESDO

The features of an app reflect the developers' vision, which includes the app's "purpose, target user base and scenarios of use" (Light, Burgess, & Duguay, 2018, p. 889).[5] One of the cofounders of Rela (called The L when it first launched in 2012), Lu Lei, told the media that he built the app for his lesbian friends:

> Although I think that my several lesbian friends are beautiful, adorable, and talented, they are still single. . . . Sometimes when we were dining

together, everyone was enjoying, and suddenly she [one of his lesbian friends] became quiet. I think that every human needs love, or at the least a company. ("The L chuangshiren," 2015, para. 9)

Back in 2012, there were no mobile dating apps for queer women in China. Queer women primarily used online forums and remained anonymous. Lu suggested that an app running on smartphones would enhance the visibility of individuals and the community as a whole ("Rongzi baiwan meijin," 2016). He also contrasted the ways queer men and women socialized. He noted that queer men's sociality was based mainly on sexual connections (similar to Gudelunas, 2012) while queer women's sociality was built on chatting, interacting, and building long-lasting relationships.

Lu translated his vision of building a queer-friendly environment into activism. On May 20, 2017, Rela, together with Qinyouhui 亲友会 (Parents, Families and Friends of Lesbians and Gays of China), launched a campaign for marriage equality. They arranged for eleven mothers of queer adult children to go to People's Park in Shanghai, which is a famous match-making market, to look for a spouse for their children (see chapter 2 for a similar park in Guangzhou). These mothers handed out informational flyers on marriage equality and LGBTQ rights to park visitors (Bhandari, 2017). The police came after an hour and ordered the parents to leave. Rela's operations were later temporarily suspended.[6]

Rela's concerns for building a queer women's community and social engagement were also reflected in the company's creation of features that were dramatically different from those of other Chinese gay dating apps. First, similar to the relationship status feature on Facebook, Rela users can "bind" their account to their partner's. The app counts how many days they have been together ("Rongzi baiwan meijin," 2016). With this feature, users can present themselves to other app users as a couple. This feature is not available on gay apps such as Blued or Aloha. Second, the landing page of Rela has an "status updates" feature. Users can see updates from people they have been following (figure 5.1). Also embedded in this page is a feature called "topics" where users can view and participate in discussions on popular subjects. The "people nearby" feature allows users to locate strangers. Although this feature is on the landing page of Blued,

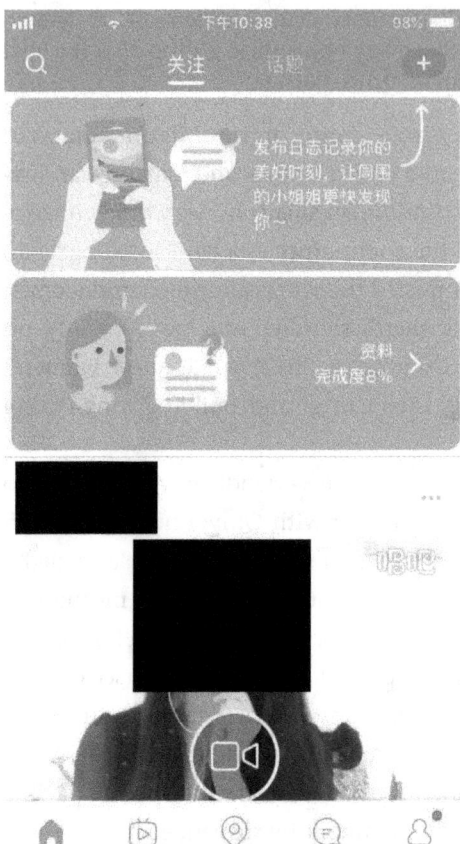

Figure 5.1
The landing page of Rela displays updates from existing connections. (Screenshot taken
by the author on March 24, 2019)

it is relatively buried in the interface of Rela, relegated to third feature in
the navigation bar. If we consider the landing page of an app to be the
most important feature envisioned by the app developer, Rela is more
geared toward maintaining existing relationships and cultivating a community than developing new dyadic relationships. Third, related to "topics," Rela launched a digital magazine called *Rela Zhoukan* 热拉周刊 (*Rela
Weekly*), which hosts original content about careers, romance, and other
issues pertinent to the community written by renowned queer writers,
bloggers, and artists ("The L chuangshiren," 2015). Lu's idea that queer

Figure 5.2
The landing page of Lesdo has a feature called "community" where users can view blog posts created by others. (Screenshot taken by the author on March 13, 2019)

women's relationships are less about casual, short-term connections than about building a long-term community is inscribed in these features.

Such an explicit focus on community-building is also apparent in Lesdo. Tingting Liu's (2017) interview with Lesdo's management shows that the app prides itself on its "community" feature. This feature also appears on the landing page of the app (figure 5.2). Users can view, like, comment on, and share blog posts written by other users. These posts consist of up to five thousand Chinese characters and cover topics such as coming-out advice, lesbian-themed movies, health information, and cooperative marriage. Liu argues that the "community" feature "highlights the possibility of providing virtual community-based care for queer subjects" (p. 303).

Such emphasis on community-building in both Rela and Lesdo is likely one of the reasons that the accounts my queer female informants provided fell into the stereotype that they were not interested in sex. As the interview of Lu Lei shows, Rela was built with the biased assumption that queer women cared about long-lasting relationships more than short-term casual relationships. So was Lesdo. But features alone do not dictate what users do. That is, having an "status updates," "topics," or "community" feature does not automatically preclude queer women from seeking sex and connect them with their community. For this reason—returning to my emphasis on perception of affordances—it is necessary to consider how users have interpreted these features.

INTERPRETATION OF RELA AND LESDO

My informants said they often developed beliefs about the dating apps they intended to use before they used them. These beliefs came from sources like marketing materials, online commentaries, and stories from their peers. For instance, Momo is known by the general public as *yuepao shenqi* 约炮神器 (a magical tool for hookups) due to a viral video and its initial advertising campaign (T. Liu, 2016). In chapters 2 and 3, I show how some Momo users were motivated by these marketing techniques, while others actively resisted them. My queer male informants held similar beliefs about Blued, for example, partly because they read numerous stories shared on online forums about Blued being a hookup app before they downloaded it.[7]

My queer female informants told me a very different story. Only a few of them had considered lesbian dating apps such as Rela and Lesdo as hookup apps before they used them. Instead, most saw these apps as a platform to make friends and obtain information about their *quanzi*. Recalling how she came to download Lesdo, Shawn said the app was recommended by her lesbian friends living in Beijing. "They said this app is great for making friends. So I was curious and wanted to find out how many people similar to me were around." This belief that lesbian dating apps are mainly for making friends, but not for hooking up, was also shared by Xiaoyan. She explained why she downloaded Rela:

> I wanted to make friends of the same sexual orientation as mine. I thought this way of chatting and interaction would be more comfortable. I did not have issues with my identity, but sometimes I thought that, as a lesbian . . . interacting with straight women is pretty different. . . . None of my friends had even used the app [for hooking up].

Apart from word of mouth, my queer female informants learned about these apps through microblogging platforms such as Douban and Weibo. Douban and Weibo are not platforms tailored to queer women, but users can follow people or organizations that are known to their *quanzi*. Shawn remembered how she first read about Rela from a story posted on the official Weibo account of Qinyouhui. The story described Rela as an information portal for queer women. Xiu, age twenty-three and lesbian-identifying, was attracted to Rela by its radio program: "At that time, Rela had a radio program that was quite appealing to me. . . . I like its topics because we do not have many radio programs for us." Although Kay Siebler (2016) laments the diminishing influence of LGBTQ organizations in the digital era in the American context, the internet is where these organizations in China have flourished.

Why did my informants view lesbian dating apps as a platform for platonic relationships and information about their *quanzi* rather than for sexual relationships? A plausible reason is the presumed relative underemphasis on hookup culture in the queer women's *quanzi* that my informants belonged to. Becky, age thirty-four and bisexual-identifying, commented on the hookup culture as follows:

> Lesbians and gay men are very different. For gay men, perhaps they will have sex after they find each other "okay." But for lesbians, we chat, chat, chat, and keep chatting. It is tiring actually. You keep chatting, understanding each other, but not for the purpose of dating. And then, [you] talk about life, ambition, and different topics. Only after all of these there is a possibility of having sex.

Becky's comment is consistent with the way Rela's founder envisioned queer women's sociality. Xi, who also identified as bisexual but was ten years younger than Becky, shared a similar view:

The prerequisite [for hookups] is that we need to be able to communicate. I remember my friends and I had a conversation before. We said, "If we meet up with someone whom we plan to hook up with but cannot talk about our philosophy of life, why should we have sex?" I think a sexual relationship is also communication. If you two have no interest in communicating, it won't be harmonious on the bed either.

My queer female informants often joked about the need to know the so-called *sanguan* 三观 (three views)—worldview, outlook on life, and values—of their potential hookup partners before hooking up. In contrast, only one of my straight female informants, Kangqi, said that *sangguan* was important to finding a casual sex partner. Further, none of my male informants, straight or queer, mentioned *sanguan* as a criterion for seeking casual sex partners. *Sanguan* remained characteristic of the expectations queer women had for relationships forged through dating apps.

In their recent research on women's motivations to have casual sex in the Canadian context, Heather Armstrong and Elke Reissing (2015) found that both straight and queer women were equally likely to consider physical appearance as a salient criterion when looking for a causal sex partner. They remark that the shifting cultural norms in Canada have allowed women to feel more comfortable seeking casual sex for purely physical reasons. In chapter 2, I mention that my straight female informants felt empowered by their pursuit of casual sex. This result is consistent with Armstrong and Reissing's study. However, such rhetoric was not found among my queer female informants. Indeed, my historical review above points out that Chinese lesbian activism has emphasized the desire to integrate into mainstream society and that queer women must display "public correctness" (Engebretsen, 2014; Kam, 2013). In this sense, the underemphasis on hookup culture in Chinese queer women's *quanzi* reflects the sexual conservativism that the members in the *quanzi* perceive to be crucial to their survival in the heteronormative society. Using the distinction I make in the previous chapter, in-app, my queer female informants conformed with an image of an ideal user who was more interested in building a community, while, out-of-app, queer women in China have never been allowed to pursue serendipitous sexual encounters.

COMMUNAL CONNECTIVITY AND ITS MULTIPLE OUTCOMES

My analysis above demonstrates two aspects of the Chinese lesbian dating app culture. On the one hand, lesbian dating apps have dominant "status updates," "topics," and "community" features. On the other hand, their users regard them as platforms for platonic friendships and information. The affordance resulting from the intersection of these two aspects is what I call *communal connectivity*. Jessica Fox and Bree Mc-Ewan (2017) refer to this as "network association," which "enables group members, no matter how disparate or geographically distant, to identify other members" (p. 303). Dating apps such as Rela and Lesdo similarly make users a "promise" (Scarantino, 2003) that if they want to, they can be connected to the larger *quanzi*. I suggest that because this affordance stresses the communal aspect of dating apps, it is fundamentally different from the five affordances I identify earlier.

Evans et al. (2017) propose a three-pronged conceptual framework to identify what property of communication technology can be considered an affordance. Their first criterion is that it cannot be a feature. Communal connectivity is not characteristic of a single feature of dating apps. The features contributing to communal connectivity are the "Status updates," "Topics," and "Community" features or, on a more basic level, the mobile data transmission. Second, an affordance has range. Communal connectivity is neither present nor absent but encompasses a range of engagements. It makes sense to say "User A is fully connected to the *quanzi*" or "User B is somewhat connected to the *quanzi*" based on how frequently they post, read, or comment on a blog post, for example.

Having passed two of the three thresholds, let us consider the third criterion Evans et al. (2017) propose: an affordance is not a consequence of using the technology. They remind us that "an affordance can be associated with *multiple* outcomes" (p. 40, emphasis in original). Based on this criterion, they argue that anonymity is an affordance, but privacy is not. Privacy "is an outcome resulting from affordances such as visibility or editability" (p. 44). I agree that an affordance should be able to lead to multiple outcomes. For example, James Gibson (1979), who coined the term *affordance*, points out that pressing the blade of a knife against something can result in either cutting or hurting. However, I disagree with the reasoning offered by Evans et al. because their argument is reductive. They assume there are some elementary affordances that produce various outcomes. As privacy is the result of visibility and editability, it cannot be an affordance itself. What Evans et al. have failed to note is that anonymity, which they consider an affordance, also comes from visibility and editability. Anonymity can be considered an outcome when people choose to hide their identity from others (that is, visibility) or delete their information (that is, editability). In fact, with this reductionist logic, even editability cannot be an affordance because it results from the affordances that we can input, store, retrieve, and replace data in a digital network.

I suggest setting aside the question of whether a property is a result of other affordances. Instead, we should focus on whether that property can lead to different outcomes. If multiple outcomes are possible, then it passes the third threshold. My modification to the criterion of Evans et al. (2017) highlights the contingent nature of affordances. Although communal connectivity is a result of mobility and visibility, it is primarily about connecting to the larger *quanzi* rather than to individuals. My research demonstrates that connecting to one's *quanzi* leads to multiple outcomes, including emotional support, information, and advocacy.

My informants' narratives prominently described using dating apps for emotional support. Similar to the queer male informants discussed in the last chapter, my queer female informants faced tremendous social pressure. Most of them had not come out to their parents and wanted to be filial daughters. Being a filial daughter involves making the right choices so that one's parents will not worry about one's life (Eklund, 2018). Many of my informants knew that their parents would be extremely worried about their future if they came out. Therefore, they hid their sexual orientation from them. In addition, my queer female informants working in government-related institutions also were greatly concerned about the visibility of their sexuality. For instance, Shawn worked as a branding executive in a state-owned enterprise. From the beginning, she knew she had to be very careful about not revealing her sexual orientation to her colleagues. Later, she found out that her department supervisor was gay but had been living a double life in a heterosexual marriage. This further reinforced her belief that she must keep her sexual orientation a secret at work.

With communal connectivity, dating apps enable their users to seek emotional support outside of the sphere of work and family. This chapter begins with a quote from Xiaoyan. She said that lesbian apps helped their users to know that "they are not alone in this world." Xi added, "Everyone shares the same worry but with subtle differences. I think it is very important to communicate my feelings." Dating apps, to her, were an important platform for vetting and sharing secrets. Her use of dating apps, therefore, echoes the longstanding use of the internet for the queer communities to search for belonging and identities (Campbell, 2004; Gray, 2009; Gross & Woods, 1999; Mowlabocus, 2010).

In her study of the app Butterfly, Christine Choy (2018) found that queer women in Hong Kong used forums to find information on lesbian-friendly commercial venues. Similarly, my informants sought out news and information disseminated through the blog posts and live streaming on Rela and Lesdo. One type of information described how to come out to one's parents. Charlie, age twenty-four and lesbian-identifying, recalled watching live streaming by several hosts from Qinyouhui on Rela both before and after she came out to her mother:

> Before I came out, I would watch the live streaming by Qinyouhui. I watched some lesbians recounting their experiences of coming out. I would ask them questions. Even after I had come out to my mother, I watched their live streaming where they invited some parents. I would ask these parents, "Will my mom feel bad? Will she feel uncomfortable?"

Information like this is not available from the Chinese mainstream media. This is why dating apps are important to the lives of queer women. My informants also looked for other kinds of information. For instance, Shushu, age twenty-six and identifying as queer, and her partner wanted to know how other lesbian couples lived their lives. Recently, marrying overseas has become a trend for Chinese LGBTQ people who have the requisite financial resources. Dada, age twenty-five and identifying as bisexual, recalled watching a live stream on Rela where the host meticulously described the procedure and costs. Dada also enjoyed listening to the gossip of famous live streamers. As Max Gluckman (1963) puts it, gossips "maintain the unity, morals and values of social groups" (p. 308) and an "outsider cannot join in gossip" (p. 312). To Dada, participating in gossip allowed her to imagine a group of invisible audience who she knew were somewhere in the city, sharing her interests, and most important, her sexual orientation.

Communal connectivity not only enables app users to benefit from the *quanzi* but also allows them to contribute to it. One informant, Xi, a professional psychological counselor, provided counseling on dating apps by writing blog posts and live streaming. She said,

> On Lesdo and Lespark, because there is a forum section, I can do some advocacy work, such as anti-domestic violence and anti-sexual harassment. . . . I usually start a blog post, others will reply, and then it will become a group chat. I have done live streaming as well.

Gaining a sense of belonging and membership to a larger community is a powerful drive. Research in the Hong Kong context has found that feeling of being a part of the queer community mitigates the stigma of being a sexual minority, which in turn relates to better mental health (Chong, Zhang, Mak, & Pang, 2015). These three uses of dating apps afforded by communal connectivity—emotional support, information, and advocacy—go beyond dyadic relationships. They are about seeking and maintaining the queer women's *quanzi*, the communal aspect of dating apps that was completely missing in the narratives of my straight informants and much weaker among my queer male informants.

REPRODUCING HETERONORMATIVITY

Communities enforce norms (Blackshaw, 2010). As Judith Butler (2004) writes, "norms may or may not be explicit, and when they operate as the normalizing principle in social practice, they usually remain implicit, difficult to read, discernible most clearly and dramatically in the effects that they produce" (p. 41). In certain lesbian online chat rooms, regulars actively evaluate whether newcomers are "authentic" lesbians based on subjective and arbitrary criteria (Poster, 2002). In Western gay dating contexts, the expression "no fats, femmes, or Asians," signifies the fatphobia, femmephobia, and racism that are prevalent in queer male communities (Ayres, 1999; Conte, 2017; X. Liu, 2015). The privilege given to masculine, straight-acting bodies is one of the many manifestations of homonormativity, which Lisa Duggan (2003) defines as "a politics that does not contest dominant heteronormative assumptions and institutions, but upholds and sustains them" (p. 50). The lesbian dating apps I analyze in this chapter also reproduced heteronormativity in the Chinese queer women's *quanzi*. Specifically, the videos produced by these apps and the designs of the apps assert a distinct gender role difference between the butch and the femme, or what my informants called *T* and *P*.[8]

Rela and Lesdo produced several queer-themed microfilms and online videos. Jia Tan (2018) calls these "social app videos." Most of these videos last for about ten minutes, although some are longer.[9] They generally cover the everyday issues related to being a queer woman in China, such as coming out to parents and forming romantic relationships. For example, *The L Bang* 热拉帮, a four-episode sitcom produced by Rela in 2015, describes the lives of five lesbian, gay, and bisexual neighbors in a high-rise residential building in contemporary Shanghai. The female characters find each other on Rela, and two of them later become a couple.[10] The portrayal of female same-sex intimacy in these videos is a milestone in the history of media representation. As I mention above, movies about female same-sex intimacy have been scant in China. Even when female same-sex intimacy is portrayed, it is often presented as the character's past experience (Martin, 2010). The "platform presentism" of these videos produced by Rela and Lesdo (Tan, 2018) forges a tighter temporal connection with the audience.

However, images of both androgynous or masculine lesbians and feminine lesbians are part of these videos. In *The L Bang*, Cooka and Anda have very short hair and often wear shirts, while Nana has longer hair and wears earrings and skirts (figure 5.3). Both Cooka and Anda are fond of Nana, who eventually chooses Cooka. This T/P relationship perpetuates the categorical gender binary from heteronormativity where each partner represents a masculine or feminine ideal.

If we believe the representations above reflect the vision of the app developers, heteronormativity has also been inscribed into the apps through their design. Both Rela and Lesdo probe users to indicate their gender roles. On Rela, users are offered six options on the registration page: T, P, H (versatile), Bisexual, Other, and Not disclosed. On Lesdo, only the first five options are available. Users can also use these labels to screen out unsuitable partners. Some of my informants regarded this classification as useful. Charlie, identifying as H, explained,

Figure 5.3
A scene from *The L Bang*, a queer-themed sit-com produced by Rela. From left to right are Nana, Cooka, and Anda (facing away from camera). (Screenshot taken by the author on June 18, 2019)

Well, for those who dress up more like a man, they are very clear about what they like, that is, "I just like a girl with long hair, who wears a skirt." So they will explicitly state, "I want to find a P." . . . With these labels, people can tell how you look like immediately.

The rationale Charlie offered in support of the labeling system was that labels help people manage expectations. However, my other informants said that such systems were outdated. Shushu remarked that "it is unnecessary to use a label to restrict myself." She preferred to identify as queer—a more fluid term than *lesbian*—and resisted the labels of T and P. Magda, age nineteen and preferring to be identified as a female homosexual, found Rela's emphasis on role distinction to be extremely heavy and was very "uncomfortable" with it. She connected the app design to one of the videos Rela had produced:

I did not watch it. But just based on its description, I can feel that it is putting female homosexuality into a heterosexual frame. I feel that the female lead is just biologically a female but acts the same as the men in popular TV drama.

The affordance of communal connectivity of Rela and Lesdo enabled my informants to seek emotional support, look for information, and engage in advocacy that was beneficial to their *quanzi*. But connecting with their community also meant complying with norms encoded into the apps.[11] These lesbian dating apps reproduced heteronormativity by perpetuating stereotypical gender roles in their videos and foregrounding gender role labels on the app. Although the butch-femme classification has a very long history within Western lesbian communities (Kennedy & Davis, 1993), recent research in the United States has revealed a deep divide between those who endorse these labels and those who do not. Esther Rothblum (2010) describes this as follows:

> At one end of the continuum are lesbian, bisexual or queer women who perceive butch or femme to be core identities, equal in salience to gender, race, or sexuality, and who regard these concepts as extremely important. At the other end of continuum are women who find the terms outdated or meaningless, or who embrace the terms but find that both or neither fit them well, or who are creating their own terms and definitions. (p. 41)

I respect my informants who found these labels useful. However, I would argue that inscribing these categorical labels onto the app's design assumes that someone has a fixed gender identity and can have only one gender identity at a time. Every time these labels are used, gender is reified. Reflecting on her experience of not being recognized as a lesbian by another lesbian woman, Robbin Vannewkirk (2006) argues that "attempts to read a vibe . . . must take into consideration the possibility that personal reality can shift" (p. 84). That is, gender should not be seen as an essential attribute. Critiquing a similar design on Butterfly, Denise Tang (2015) warns that "the reinforced label selection on using the app might have curbed imaginative connections" (p. 270). Codified gender roles perpetuated by these Chinese lesbian dating apps preclude change, fluidity, and imagination. To bring forth a more inclusive and flexible *quanzi*, these apps must stop constructing categorical gender differences.

CONCLUSION

A queer utopia, as envisioned by Jose Muñoz (2009), is "a space outside of heteronormativity" that "permits us to conceptualize new worlds" (p. 35). We can think of the queer women's *quanzi* supported by dating apps such as Rela and Lesdo as a manifestation of a queer utopia. Connecting with one's *quanzi* is especially important to queer Chinese women because female same-sex intimacy has been historically suppressed, and their everyday mobility has been greatly limited under a patriarchal society. Engebretsen's (2014) research on the politics of community in lesbian activism and Kam's (2013) idea of "public correctness" for individuals have revealed that in facing the double predicament of being both queer and female, these women must act carefully to survive. Perhaps because of these historical and social circumstances, the sense of having a *quanzi* has been paramount to living as a queer woman in China.

Prior research on dating apps examined the extent these apps have destroyed or reinforced LGBTQ communities (Baudinette, 2019; Choy, 2018; Duguay, 2019; Miles, 2017). Dating apps such as Rela and Lesdo, with their affordance of communal connectivity, enabled my informants to secure emotional support, look for information that was relevant to their lives, and contribute to their *quanzi*. At the same time, connecting to a community demanded compliance with gender norms, as heteronormativity was reintroduced by the apps through their categorical gender role distinction. In this light, the space of this particular set of networked sexual publics also presented a challenge to establishing a queer utopia. As a Chinese idiom says, "The water that bears the boat is the same that swallows it up." The influence of dating apps on LGBQT communities depends not only on the features of the apps but also on how users interpret these apps. The usefulness of the concept of affordance lies exactly in its ability to capture such contingencies.[12]

My analyses in the last two chapters have demonstrated that queer politics and the use of dating apps are mutually constitutive. On the one hand, the affects derived from living daily life as a queer man in China have influenced their perception of dating apps. On the other hand, the use of dating apps by queer women has helped to build a queer, albeit restrictive, *quanzi*. In the concluding chapter, I bring back the experiences of my straight male and female informants into view to return to my concept of networked sexual publics.

6

CONCLUSION: EMERGENCE OF
NETWORKED SEXUAL PUBLICS

In this book's introduction, I discuss two traditions of dating app research. One tradition considers the impacts of dating apps on influencing interpersonal processes such as relational development and self-presentation. The second tradition—which I follow in this book—is concerned with the political implications of this emerging technology. I believe that in order to appreciate and evaluate the role dating apps have in our lives, one must delve into the terrain of gender and queer politics, exposing the sociopolitical tensions and ripples experienced by dating app users. I am interested in analyzing the shared and contrasting experiences of female and male, straight and queer dating app users. To follow this path, I have proposed the concept of networked sexual publics as a theoretical lens for looking at the emerging dating app culture in China.

Early on, I provisionally defined the term *networked sexual publics* as both the assemblage of people united by their shared position in the patriarchal and heteronormative world connected by dating apps *and* the space offering a multiplicity of interpretations and relationships for the publics. Networked sexual publics consist not only of people whose objects of romantic and sexual desires are same-gender bodies. They also include people who reject the normative pattern of dating, marriage, and reproduction (Halberstam, 2005). The "publics" in networked sexual publics are therefore both phenomenological and imaginary (Hjorth &

Arnold, 2013). That is, on the one hand, phenomenological publics are forged through a wide range of networked interactions. They are forged by seeing each other on dating apps, chatting with each other, hanging out, hooking up, and developing friendships. On the other hand, its inescapably imaginary dimension emerges when people encounter challenges in using these apps because they are called to imagine the existence of people with similar gender and sexual identities who are facing the same difficulties. These challenges include, for straight women, sexual harassment and, for queer people, the fear of being outed. Networked sexual publics thus consist of people who share the same position in the patriarchal and heteronormative world coming together through dating apps.

In this book, I have directed my attention to how my informants assigned meanings to the apps they used. In the introductory chapter, I point out that dating apps such as Momo and Blued have often been viewed too simplistically as merely hookup apps. My in-depth interviews revealed more diverse sets of interpretations. I found that dating apps are not only a laboratory for sexual experiments: they are a springboard to romance and marriage, a third place between home and the workplace, a gateway to new worlds, a gallery of good-looking people, and a portal to look for business partners. Especially for queer app users, they are a potent symbol of social legitimacy and a platform for finding emotional support and information from their community. These multiform interpretations, in turn, result in various relationships—long-term romantic relationships, short-term open relationships, community-based engagement, or merely the act of gazing at people. Networked sexual publics support multiple interpretations and diverse relationships.

The concept of networked sexual publics is meant to shift our attention from technological artifacts to the political and social implications surrounding their use. Although my research was conducted in southern China, it provides insights into this global emerging culture of dating apps. In this concluding chapter, I address this question: how can we study networked sexual publics in a global context? Based on the research presented in this book and my reflections on some of the latest developments on dating apps worldwide, I present five propositions on the defining features of networked sexual publics and ways to research this emerging phenomenon. These propositions return this concept to the currents of scholarship I relied on, charting courses for future inter-sectional, queer, and feminist scholarship on emerging communication technologies.

1 In networked sexual publics, when there is resistance, there is also dominance. We should note the various manifestations of resistance and dominance.

One of the three theoretical traditions where I situated my research is the body of literature that examines the relationships between gender and technology *and* between queerness and technology, respectively. While there have been recent attempts to reconcile feminist and queer concerns (Burgess, Cassidy, Duguay, & Light, 2016; Cipolla et al., 2017; Marinucci, 2010), existing scholarship on social media and dating apps has been founded on either the feminist critique of patriarchy (e.g., Wajcman, 2007; Wallis, 2013) or the queer critique of heteronormativity (e.g., Campbell, 2004; Cavalcante, 2019; Gross & Woods, 1999; Mow-labocus, 2010). I intend for my concept of networked sexual publics to be a unifying framework for the investigation of the rise of dating apps, foregrounding the common motif in the accounts of most of my informants—simultaneous resistance and dominance.

My informants often resisted patriarchy and heteronormativity in their everyday use of dating apps. In the context of the changing status of women, I explore the opportunities and challenges that dating apps such as Momo and Tantan afforded to straight women. They used dating apps to exercise their sexual agency and reverse-objectify the men they saw on dating apps. Queer men and women resisted compulsory heterosexuality in other ways. Queer men found using dating apps to be both enjoyable and even hopeful, though their experiences were also shaped by the fear evoked by contemporary Chinese queer politics. The affordance of com-munal connectivity in lesbian dating apps contributed to the making of queer women's *quanzi*, one that was comforting due to emotional sup-port and information.

Yet the liberating potential of dating apps is not absolute. Michel Fou-cault (1979) reminds us that there is a "strictly relational character of power relationships" (p. 95). This relational character has two implica-tions. First, both resistance and dominance are immanent in the same power dynamics. The site for resistance is inevitably used for the reasser-tion of dominance. Second, the characteristics of resistance that Foucault describes include a plurality of dominances. Dating apps, therefore, are where various manifestations of patriarchy and heteronormativity seep into interpersonal interactions.

Sarah Banet-Weiser (2018) shows how digital media, through affordances of connectivity and visibility, have simultaneously enabled popular feminism and also popular misogyny. Responding to her prescient bidirectional conception of power, I illustrate how straight male dating app users reclaimed their dominance through the performance of gender and the objectification of women—the exact tactics women used to resist patriarchy on dating apps. Furthermore, app companies exercise dominance through app design. Rena Bivens and Anna Shah Hoque (2018) argue that Bumble, an app that positions itself as a feminist app through the "ladies ask first" feature, has reinforced cisgender logic by offering only two gender options and allowing its users to alter their gender only once. In a similar vein, I show that lesbian dating apps perpetuate heteronormativity by reasserting a rigid gender role distinction.

I do not believe that all male dating app users or their masculine practices are misogynistic. Nor do I imply that the companies that create dating apps such as Rela and Lesdo are conspirators in heteronormativity just due to their commercial nature. I think it is futile to look for, in the words of Foucault (1979), "the headquarters that presides over its rationality" (p. 95). The hegemonic aspect of networked sexual publics will never have a single convenient scapegoat. Instead, hegemony operates through the oppressive logics of patriarchy and heteronormativity that are constantly being reproduced. Therefore, to research networked sexual publics, our critique must go beyond technical features and consider larger social, cultural, and political systems. Scholars must recognize the inherent potential for both the resistance and dominance of networked sexual publics as they research how they manifest.

2 Networked sexual publics are about multiplicities of relationships. To reword Lauren Berlant and Michael Warner's (1998) description of the queer world, networked sexual publics "include more relationships than can be mapped beyond a few reference points."[1]

I have reiterated that users of dating apps held a variety of interpretations of the apps they used. These multiform interpretations raise the point that dating apps are not just for dating. Instead, they are an infrastructure that supports multiplicities of relationships and nonrelationships—recalling some of my informants who only looked at photographs and had no intention of interacting with others. During my interviews, I noticed a term that my informants—straight and queer, male and female—often used: *mudixing* 目的性 (purposefulness). Unlike *mudi* 目的 (purpose), *mudixing* was used to refer to the singularity of any purpose. My informants used it in an entirely negative way to criticize some app users who revealed their motives too early and too bluntly. For instance, Kangqi described some men on Momo as too eager to have casual sex. "Those people's *mudixing* are too strong . . . rushing to develop casual sex relationships with you." Damon observed that on Blued, some users tried the app for a few days, disappeared, and then reappeared again. "These people have a strong *mudixing*—that is, looking for hookups. After they hook up, they need some time to 'cool down.'"[2] In other cases, my informants used *mudixing* to refer to people who used dating apps only to sell products. Bob told me that some users "have *mudixing*, not about casual sex but about—perhaps because of their jobs or being financial consultants—promoting their products and services." Dylan, who used Momo to look for business, referred to people as having *mudixing* when they were interested only in doing business on the app. The Chinese notion of *mudixing*, therefore, refers to a situation in which users reduce the multiplicity of relationships or nonrelationships afforded by dating apps into a singular goal.

A research implication of this observation is that although networked sexual publics are structured by individuals' shared position in the patriarchal and heteronormative world, they are not about sex or romance alone. Networked sexual publics must be approached as an excess, an overflow of relational possibilities afforded by dating apps. Even in the contexts of Western apps such Tinder and Grindr, which have a relatively simple interface, scholars must not assume that a unitary purpose exists among app users. They must consider how individual users interpret these apps and consider that relational goals may not be explicit or may change during the use of an app.

This aspect of networked sexual publics also suggests that any legal regulation that outlaws an entire class of behavior on dating apps deserves more critical attention. For instance, the Singaporean government recently made sending unwanted photographs of genitals illegal (A. Wong, 2019). This move will protect dating app users from sexual harassment, which induces distress and humiliation. What we must also consider, however, is the costs of these policies and laws. Regulations should both protect users and also preserve the organic, fluid, and sometimes flirtatious nature of networked sexual publics.

3   The meanings and affects that users attach to dating apps cannot be reduced to psychological motives. Intersectionality-like thinking is necessary. A dominant theoretical framework of recent dating app studies has been uses and gratifications theory (Katz, Blumler, & Gurevitch, 1973). Originally developed to understand television viewing, uses and gratifications theory was an attempt among communication scholars to move from considering the "effects" that mass media has on people to looking at what people do with mass media. This theory assumes that audiences actively select media that satisfy specific cognitive needs. The effect of television violence, for example, cannot be assessed without knowing the audience's motives for watching television.

Using uses and gratifications, multiple research teams have identified the motives for using dating apps.[3] Unsurprisingly, all of these studies have pointed out that seeking sex is one of the common reasons for using a dating app. My central point here is that it would be an ontological error to treat sex-seeking across different groups of networked sexual publics as a homogeneous "use." Consider that, in Jed Brubaker, Mike Ananny, and Kate Crawford's (2016) sociotechnical account of gay men quitting Grindr, the same act carries various social meanings. Some gay men viewed leaving Grindr as a return to the old way of meeting new people. Others regarded it as a move to a new life stage that did not rely on seeking casual sex. Still more gay men said that leaving Grindr presented an opportunity to discuss relationship goals with their partners. Likewise, in this book, sex-seeking has very different meanings for different users. For my straight male informants, seeking sex partners fulfilled their physiological need to express their masculinity. For my straight female informants, the same practice was a tool for rejecting patriarchy and discovering their sexuality. For my queer female informants, it was a form of sociality that enabled them to look for someone with the same values and worldviews. Each of these motivations differed based on the users' identity, societal position, and other contextual factors.

Besides their cognitive aspect, networked sexual publics are also affective. Zizi Papacharissi (2014) discusses political participation in affective publics. I show that, depending on where the networked sexual publics are located in the heteronormative world, using dating apps triggers and reinforces different emotions. By turning our attention to the meanings and affects people hold toward dating app use, I have been able to peel back the multiple layers of personal and political endeavors that a single behavioral motive label fails to distinguish.

Looking to future research, one way to ensure that close attention is paid to the social meanings and affects among different groups is to adopt intersectional thinking. As I demonstrate in this book, analyzing one group of users (straight women, straight men, queer men, and queer women) at a time and contextualizing their experiences with dating apps based on their positionalities in contemporary gender and queer politics in China allowed me to discover their unique struggles, dilemmas, opportunities, and challenges. Cara Wallis (2013) puts it aptly when she writes, "in this way, we can . . . extract the fine nuances and diverse shades of meaning that technologies have for different groups, thereby creating a richer, 'thicker' understanding of technology, culture, and social change" (p. 188). Achieving this thickness, a characteristic of anthropological research, can enrich future research on emerging communication technologies.

As quite a lot of dating app research has relied on survey data and statistical analyses, I also wish to address the hazards of studying networked sexual publics quantitatively. Very often, scholars, including myself, have regressed the motives for using dating apps on gender, sexual orientation, and the interaction of the two. They then removed the interaction term from the regression model if it was not statistically significant. This way of incorporating intersectionality-like thinking into research is what Ange-Marie Hancock (2016) calls the "intersectionality-as-testable explanation" approach, which she warns misrepresents intersectionality theory. Intersectionality theory offers a research paradigm that conceives of the social world and human practices as structured by multiple axes of domination. That is, intersectionality is an *a priori* assumption about the phenomena that we are researching. Accordingly, it is essential to assume "combinations of conditions as the default analytical starting point" (Ragin & Fiss, 2017, p. 11).[4] My approach taken in this book—assuming that the preconceived axes of power of patriarchy and heteronormativity have already put my informants into different sets of struggles—let me explore inductively their intersectional experiences.

4 Networked sexual publics are regionally specific. We should relate the emergence of networked sexual publics to the historical, social, and cultural environments of the region.

This study was inspired by scholarship in mobile cultures of the Asia Pacific, whereas most studies of dating apps have been conducted in Western countries. This book provides an alternative perspective by examining the dating app culture in urban China, extending scholarship on Asian mobile cultures (Berry et al., 2003; Cabañes & Uy-Tioco, 2020). The central concern of this scholarship is how to conceptualize a global phenomenon in a local context. The emergence of "Asia as method" in science and technology studies also suggests that a phenomenon in Asia should not be used as a case to "test" Western theories. Instead, it should be regarded as a constitutive element of that very phenomenon (Anderson, 2012). The fact that Western dating apps were not popular in China provided a window into the local specificities of both the apps and their users. This book has enriched our understanding of the role dating apps play in shaping gender and queer dynamics in the Chinese context through engaging with indigenous concepts and specific sociopolitical circumstances that are inescapably local. In particular, discussions in chapter 2 and 3 contribute to technofeminist scholarship where gender relations and technology use are mutually constitutive (Wajcman, 1991, 2006, 2007), and by tracing the social history of women in China and employing the indigenous concept of *wen-wu* 文武 (literary-military) masculinities, they also reveal the cultural specificity of dating apps' interpretations. For example, women's interpretation of dating apps as a springboard for marriage was intensified by the threat of being stigmatized as *shengnü* 剩女 (leftover women), a politicized and gendered pejorative. Meanwhile, men's interpretation of the very same set of apps as a platform of business was reflective of the changing idealization of *wen* masculinity in neoliberalized China.

Chapters 4 and 5, acting as a pair, bring our attention to the affect, affordances, and experiences of digital technology among queer people in a specific regional context (Campbell, 2004; Cassidy, 2018; Cavalcante, 2019; Landström, 2007; Molldrem & Thakor, 2017; Mowlabocus, 2010). I demonstrate that stagnant Chinese queer politics have limited the affective experience of using dating apps and have shown that the rhetoric of pride so ingrained in the Western LGBTQ movement does not apply to Chinese queer politics. The double predicament experienced by Chinese queer women rendered their apps a platform predominantly for community-building.

Therefore, future scholars who want to understand the networked sexual publics in another region must first be familiar with local culture and society. For instance, Jason Vincent Cabañes and Christianne Collantes's (2020) work on Filipino female dating app users living in Manila invokes the concept of "digital flyovers" to describe the way dating apps have allowed these women to reach out to foreigners, bypassing local Filipino men whom they deem uncosmopolitan. They conclude that women's desire for foreigners must be understood against the fraught colonial history of the Philippines.

Knowing the local language, in additional to culture, can further offer an edge in one's scholarly analysis. For example, Larissa Hjorth (2003) uses the Japanese concept *ma* 間 to examine mobile phone culture in Tokyo. The word *ma* means a pause in time or space. It is used in phrases such as *mamonaku* 間も無く, which literally means "pause no more" or, conceptually, "shortly." She appropriates the term to refer a conceptual ambiguity that suggests both presence and absence simultaneously. On public transport in Japan, it is considered rude to talk on one's phone or speak aloud. Therefore, most communications are conducted through texting. *Ma* captures a space where no communication and full communication happen at the same time. This example illustrates that culturally accurate and localized readings are most palpable when indigenous concepts serve as a theoretical lens to unpack local phenomena. What kinds of indigenous terms do people from different regions use to refer to dating apps? What are the connotations of those terms? What cultural repertoires do the terms draw from when people use them to name dating apps?

5   There is a limit to networked sexual publics. Not every individual has access to networked sexual publics, and for those who do, there is a price to being connected.

Networked sexual publics have opened up new possibilities for socializing, relationship-seeking, sexual experimentation, and business practices. This is a departure from the restrictive modes of intimacy and connectivity that were available to earlier generations of Chinese. However, networked sexual publics are predicated on the availability of communication technologies and network services. For this reason, they are not equally accessible to all. The "information haves" enjoy seamless connections to networked sexual publics using the latest smartphones and unlimited mobile data plans. The "information have-less"—connected by low-end or used smartphones and public free Wi-Fi—enjoy only intermittent access to networked sexual publics (see J. L. Qiu, 2009). The "information have-nots," unfortunately, resort to other, premediated forms of affective infrastructure. For instance, mobility may be hampered if a person can use only the free public Wi-Fi that is available only at certain locations and certain times. Such an economic and informational hierarchy reminds us that, although the features of a dating app may remain identical across the globe, its affordances may not be so homogeneous.

Apart from accessibility, one often has to pay to access networked sexual publics. Dating apps are a capitalist product. To sustain their operation, dating app companies generate revenue from their users. Some apps offer premium services to those who pay a monthly fee. For instance, on the gay dating app Aloha, users with premium subscriptions can search for people based on location. Monthly subscriptions are often renewed through auto pay, providing the app companies with a stable income stream. These relationships of data and money can endure, surprising even the users who signed up for the service. One of my straight female informants, Fanny, told me that she had forgotten she had subscribed to Momo through auto payments until she noticed that money was being withdrawn from her Alipay account.[5] Other apps, like Momo and Blued, rely heavily on live streaming as an income source (Deng, 2018; Edmunds,

2017). Viewers purchase digital gifts through the apps and send them to their favorite live streamers. In the process, the apps take a share of the gifts' monetary value. Only three of my sixty-nine informants reported buying and giving digital gifts, and they spent from CNY100 (~USD15) to CNY1,000 (~USD145). Dada shared her experience. "I really wanted to listen to her [her favorite live streamer] sharing, know her stories. She invited us to give her gifts, so I did!" In recounting their experiences of buying digital gifts, my informants focused solely on their appreciation of the live streamers, without paying attention to how their gifts had contributed to the app companies' business.

Those who do not pay the app companies in money provide their data in exchange for the "free" services. Dating services companies are data companies. Kath Albury, Jean Burgess, Ben Light, Kane Race, and Rowan Wilken (2017) describe the ways personal data are produced, stored, and capitalized in what they call the "data cultures" of dating and hookup apps. Individuals are often unaware of how much data a single app has collected from them (Duportail, 2017). The potential for aggregating data to make even greater profits is massive for groups like Match Group, which owns dozens of dating services, including Match.com, OkCupid, Tinder, and Plenty of Fish.

These capitalist data cultures also align uncomfortably with China's policies on surveillance: "For their part, digital companies are rarely willing to discuss the details of law enforcement and intelligence agencies' access to their customer databases, or the degree to which they assist or resist such access" (Albury et al., 2017, p. 8). This is why the American government is worried that Grindr, which is still owned by a Chinese company at the time of writing, would surrender its users' data to the Chinese government (Sanger, 2019). Among my informants, however, only a few had ever thought about data privacy. Most did not know that the apps they were using could sell their data to third-party advertisers or give them to the government. Some did not even think their data were valuable. Others did not think they could escape from the data cultures and surveillance, given that their lives were based so much on the Chinese internet and the dozens of apps on their phone. For these reasons, scholars must consider the cost of being networked and understand that due to their position in the economic and informational hierarchy, not everyone has equal access to networked sexual publics. Although issues of access and surveillance are not apparently addressed by the theoretical foundations that I relied on, these are crucial areas for further research.

*   *   *

In this book, I have provided a substantial discussion of the experiences of men and women in China with different sexual orientations. By doing so, I have painted a more comprehensive picture of the gender and queer politics of dating apps in China than has been available to date. My concept of networked sexual publics not only captures the defining characteristic of the Chinese dating app culture, but it also provides guidelines for future research. Networked sexual publics involve resistance and dominance; they are about multiple users having multiple interpretations pursuing multiple relationships; they have a strong regional specificity that requires in-depth knowledge of the culture and society of the region; and they involve costs that restrict access.

To return to the story of Nancy: At the time of writing this conclusion, I contacted her again via WeChat to see if she had tried the lesbian dating apps I had recommended to her earlier. I wanted her answer to be, "Yes, I did. I am going on dates with a girl from my hometown!" I wondered how persuasive this book would be if I had demonstrated the disruptive potential of dating apps in today's Chinese patriarchal and heteronormative society. Her experiences would then demonstrate how networked sexual publics could open space for people to reach outside of heteronormative society. Her answer, however, was not what I had hoped for. She replied, "I think I still, comparatively speaking, prefer men." I was disappointed for a second, but then I started to think: wasn't the fact that Nancy could have had a fleeting thought to question her sexual orientation and that we could openly discuss this issue already indicative of the tensions and ripples created in the old system by this emerging new communication technology? If networked sexual publics are anything, they are about potential—the potential to discover oneself, to bring forth a more gender-equal and queer-friendly world, and to imagine a new form of intimacy and sociality.

APPENDIX: METHODOLOGICAL REFLECTION

In this methodological reflection, I provide details of my field trips and elaborate on the methodological and ethical challenges I encountered in the process of recruiting informants and interviewing.

FIELD TRIPS

I conducted two separate field trips to Guangzhou to collect data for this book. The first field trip took place from August to December 2016. All my interviews with straight dating app users were conducted in this field trip. The second field trip, focusing on queer dating app users, was conducted from July to September 2018.

CHALLENGES IN RECRUITMENT

Recruiting informants for this study was not an easy task. Thinking that there was no better way to contact dating app users than reaching out to them via the apps, during my first field trip, I first set up a "researcher's profile" on two mainstream dating apps, Momo and Tantan. During my second field trip, I did the same on two popular gay dating apps, Blued and Aloha. I crafted the profile so that my academic identity and the purpose of the research were foregrounded (see figure A.1). On Momo and Blued, which do not require mutual liking to start a conversation, I sent out one round of recruitment messages to people near me, containing the purpose of my research. The responses to my initial recruitment messages fell into three categories. Some users were interested in my research and signed up for it. Others stopped replying to me after learning more about the research. One particular user from Momo accused me of using "academic research" as a pretense for *yuepao* 约炮 (hooking up). I apologized and ceased to contact her. Still others, which were the majority, did not reply to my message at all. Considering that typical dating app users might not expect to be approached by researchers on the apps,

Figure A.1
This was my researcher's profile on Tantan. Left: Following my name "Sam Chan" is a phrase indicating I was on the app for conducting my dissertation research. Right: I explain my intention to look for interview informants and list the criteria for participating. (Screenshot taken by the author on November 28, 2016)

I did not want to cause too much disturbance for them, and I decided not to send any follow-up messages. At the same time, on Momo and Blued, I regularly posted via the "status updates" feature—which contained my research purpose—so that the users around me could become aware of my project without my having to send unsolicited messages to them. I waited for people to contact me. This passive method attracted only one person who volunteered to participate in my research.

I suspected the low participation rate from the methods above was due to the one-way push of the recruitment messages. Looking for research informants is similar to looking for a romantic partner on dating apps—both require mutual interest. Therefore, I switched my recruitment platform to Tantan and Aloha, which have the "swipe" feature. I basically liked every profile that appeared on my app and sent out a recruitment message to those who also liked my profile, presuming that those who liked my profile were interested in my research after reading it. Through these platforms, I successfully recruited around half of the straight female informants and some queer male informants.

Second, I attended a public lecture on women's sexuality given by Dr. Pei Yuxin from Sun Yat-sen University, Guangzhou, and I recruited participants from the audience. Dr. Pei is a sociologist of gender and sexuality studies. She was kind enough to introduce me to her audience and allow me to hand out leaflets on my research. In this way, I recruited another half of the straight women and some straight men.

Because Guangzhou has an active LGBTQ support network, my third channel for recruitment was through LGBTQ organizations. I contacted the person in charge at Tongcheng 同城 (the Gay and Lesbian Campus Association in China) and Qinyouhui 亲友会 (Parents, Families, and Friends of Lesbians and Gays of China), respectively. They helped send out recruitment messages to their members and volunteers. I recruited half of my queer male informants and most of my queer female informants in this way. I did not register the accounts on any lesbian apps because these apps explicitly prohibit men from joining them (see figure A.2).

Finally, many informants, after our interviews, were willing to invite their friends to participate in the study. These informants sent out invitations, which I drafted, to their friends. I waited for responses. Most of my

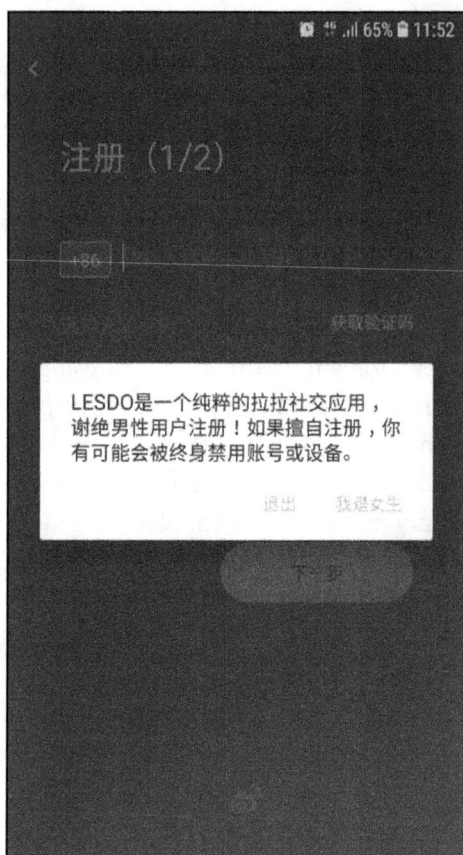

Figure A.2
Lesdo, a dating app tailored for queer women, explicitly states that the app is only for *lala* 拉拉 (lesbians) and does not welcome any male users. (Screenshot taken by the author on July 18, 2018)

straight male informants were recruited in this way. To protect the informants' identities, I did not cross-check who had referred whom.

In total, I interviewed sixty-nine dating app users. While there is no a one-size-fits-all rule for the number of informants needed in a qualitative study, researchers have pointed out that the number depends on factors such as the homogeneity of the group (Guest, Bunce, & Johnson, 2006) and saturation of knowledge (Bertaux, 1981). Within each group, my informants were relatively homogeneous: all of them identified as ethically Chinese, except one informant, Anthony, was three-quarters Chinese and one-quarter Indian but had lived his life completely as a Chinese; they were living in the same region; and most of them were born in the 1980s and 1990s. Further, I was able to recognize some patterns in my informants' narratives and experiences after conducting a dozen of interviews within each group. My time in city also allowed me to conduct additional interviews to ensure knowledge saturation.

Concerning the recruitment and sampling processes, I have two things to note. First, my experience of recruiting people for my research demonstrated that trust and rapport are crucial. On Momo and Blued, sending unsolicited messages to strangers who had no prior relationship with me barely helped me get them on board. However, people responded fairly positively when they were referred by either friends or organizations that they trusted or when they were interested in my research after reading my profile on Tantan and Aloha. Second, qualitative research like mine that is based on convenience or snowball sampling does not look for observations that can be generalized to the entire dating app user population. Instead, my informants' experiences with dating apps offered me insight so I could theorize the gender and sexual dynamics in the digital context.

ETHICS IN INTERVIEWS

As I point out in the main text, interviewing is a social process. Understanding that the interviews could involve personal and socially sensitive issues, the informants also determined the venues where they wanted to meet. Most people picked a public yet quiet coffee shop or restaurant. Some decided to use their offices.

Interviewing also involves a power relationship between the researcher and the informant. How to present myself became the first major concern in conducting the interviews. I selectively highlighted different aspects of my identity to make my informants feel more comfortable sharing their views and experiences related to sex and love online. Michael Schwalbe and Michelle Wolkomir (2003) remind us that, to a male informant, "an interview situation is both an opportunity for signifying masculinity and a peculiar type of encounter in which masculinity is threatened" (p. 57). They argue that the nature of academic interviewing—requiring the informants to follow the lead from the researcher, to self-disclose, and to reflect on their own past behavior—often puts men in an uncomfortable position. So to men, I often said, "Since I don't know how people behave in China, because people in Hong Kong or the United States are different, why don't you tell me. . . ." By foregrounding my foreignness, I let my informants play the role of "experts," giving them some control over the conversation.

Shulamit Reinharz and Susan Chase (2003) are concerned with the power relationship between a male researcher and a female informant, especially when gendered experiences were the subject. Likewise, Travis Kong, Dan Mahoney, and Ken Plummer (2003) raise a similar concern regarding straight researchers and LGBTQ informants. Some feminists have suggested that researchers self-disclose to reduce the distance between themselves and their informants. I followed this practice by sharing my experiences on dating apps with my informants. Whenever it felt appropriate, I foregrounded my sexual orientation to my straight female informants and my queer informants. Although Hongwei Bao (2018) describes Guangzhou as "one of the most queer-friendly cities in China" (p. 15), given that being gay is still not totally accepted by the majority of Chinese (Bao, 2018; Kong, 2011; Zheng, 2015), I was not comfortable disclosing my sexual orientation to my straight male informants. I was afraid that this would turn them away or damage my authority. To them, I focused on myself as a man. Whenever these male informants asked about my hookup and romance history, I shared with them without explicitly mentioning the gender of my partners by using the word keoi 佢, which is a gender-neutral pronoun in Cantonese, or the word ta, which can be interpreted as either 他 (he) or 她 (she) in Putonghua.

The second concern I want to point out is the nature of my interviews. I would describe my interviews as ethnographic, aimed at revealing the emic views of dating app users. However, my research is not ethnography. Rooted in anthropology, ethnography requires researchers to immerse themselves in the everyday lives of their informants. Some anthropologists even advocate experiencing what their informants normally do, including having sex with them if the research topic is related to sexuality (McLelland, 2002). Esther Newton's (1993) discussion of the erotic equation in ethnography reveals the presence of intimate relationships between researchers and their informants. I studied the experience of using dating apps, but I did not engage in in-app interactions with people I met on these apps beyond the recruitment stage. This was my deliberate choice because I believed that experiencing what my informants did, such as flirting on these apps, would jeopardize my professional identity and create inappropriate expectations on either side (Boellstorff, Nardi, Pearce, & Taylor, 2012). Therefore, I was very careful to indicate to my straight female informants and queer male informants that our meeting would be completely professional, and I foregrounded my researcher's identity in both the online conversations and face-to-face encounters.

DETAILS OF INFORMANTS

Table A.1 provides details of the sixty-nine informants who generously shared their experiences with me.

Table A.1 Details of the informants at the time of interview

Name	Age	App used[a]	Relationship status	Sexual identity (applicable to queer informants)	Education	Occupation	Monthly income[b]
Straight women (interviewed in 2016)							
Rosy	21	Tantan	Single, never married		Secondary school	Undergraduate student	Low
Chloe	23	Momo, Tantan	Dating		Bachelor's	Graduate student	Low
Brady	23	Tantan	Single, never married		Associate's	Interior designer	Midrange
Xiaolan	23	Tantan	Having a close emotional and sexual partner overseas but not defined as a boyfriend		Bachelor's	Graduate student	Low
Amanda	24	Momo, Tantan	Single, never married		Bachelor's	Graduate student	Low
Queenie	25	Tantan	Single, never married		Bachelor's	Studio assistant	Midrange
Nikki	26	Tantan	Dating		Secondary school	Salesperson	Midrange
Jessica	27	Momo	Single, never married		Bachelor's	Civil servant	High
Nancy	28	Tantan	Single, never married		Secondary school	Internet merchant	Midrange

Table A.1 Continued

Name	Age	App used[a]	Relationship status	Sexual identity (applicable to queer informants)	Education	Occupation	Monthly income[b]
Xiaoshan	29	Momo	Married		Secondary school	Masseuse	High
Polly	29	Momo	Single, never married		Bachelor's	Translator	Midrange
Xiaojiao	30	Momo, Tantan	Single, never married		Bachelor's	Immigration consultant	Unstable
Fanny	31	Momo, Tantan	Single, never married		Bachelor's	Administrative assistant	Midrange
Katie	33	OkCupid, Tinder	Single, never married		Master's	Civil servant	Midrange
Kangqi	34	Tantan, OkCupid	Open relationship		Bachelor's	Writer	Unstable
Coco	34	Tantan	Married		Bachelor's	Garage owner	Midrange
Jennifer	37	Momo, Tantan	Single, divorced		Bachelor's	Administrative assistant	Midrange
Yiping	38	Momo	Married		Bachelor's	Doctor	High
Wenwei	c	Tantan	Single, divorced		Bachelor's	Financial consultant	c

Continued

Table A.1 Continued

Name	Age	App used[a]	Relationship status	Sexual identity (applicable to queer informants)	Education	Occupation	Monthly income[b]
Straight men (interviewed in 2016)							
Xiaoli	19	Momo, Tantan	Single, never married		Secondary school	Undergraduate student	Low
Xiaolong	20	Momo	Single, never married		Associate's	Exhibition worker	Midrange
Roy	21	Momo, Tantan	Single, never married		Secondary school	Watchmaking apprentice	Low
Taibai	24	Momo	Dating		Bachelor's	Trade executive	Midrange
Nathan	25	Momo, Tantan	Single, never married		Master's	Marketing executive	High
Fred	25	Momo, Tantan	Dating		Associate's	Tourism officer	Midrange
Anthony	28	Momo, Tantan	Married		Secondary school	Theme park manager	High
Jiazhi	28	Tantan	Single, never married		Bachelor's	Office executive	Midrange
Clement	28	Momo, Tantan	Dating		Bachelor's	Bank officer	Midrange

Table A.1 Continued

Name	Age	App used[a]	Relationship status	Sexual identity (applicable to queer informants)	Education	Occupation	Monthly income[b]
Dylan	29	Momo, Tantan	Single, never married		Associate's	Electronic developer	High
Victor	30	Momo	Dating		Bachelor's	Civil engineer	Midrange
Eric	31	Momo, Tantan	Married		Secondary school	Vehicle technician	Midrange
Alan	32	Momo	Dating		Master's	Civil servant	High
John	36	Tantan	Married		Doctorate	Lecturer	Midrange
Fung	36	Tantan	Single, divorced		Bachelor's	Café owner	Midrange
Bob	37	Momo	Single, divorced		Bachelor's	Property manager	Midrange
Queer men (interviewed in 2018)							
Chung	19	Aloha	Single	Male *tongzhi*	Secondary school	Undergraduate student	Low
Xiaomao	19	Blued, Aloha	Single	Gay	Secondary school	Undergraduate student	Low

Continued

Table A.1 Continued

Name	Age	App used[a]	Relationship status	Sexual identity (applicable to queer informants)	Education	Occupation	Monthly income[b]
Xiaoming	21	Blued, Aloha	Single	*Tongzhi*	Secondary school	Undergraduate student	Low
Ginger	22	Aloha	Having a "weekend lover"	Gay	Bachelor's	Photographer	Low
Rice	23	Aloha	Single	Gay	Bachelor's	Lawyer	Low
Gui	23	Blued, Aloha, Tantan	Having a regular sex partner	Queer	Bachelor's	Unemployed	Low
Allen	24	Blued, Tantan	Single	Gay	Bachelor's	Health and safety officer	Midrange
Tesla	24	Blued, Aloha	Single	Gay	Bachelor's	Software engineer	Midrange
Green	24	Blued, Aloha	Single	Gay	Associate's	Communication executive	Low
Ezra	24	Blued, Aloha, Fanbaizi, Tantan, Momo	Single	A man who desires another man	Bachelor's	Civil engineer	Low
Damon	25	Aloha	Single	Gay	Bachelor's	Market researcher	Midrange

Table A.1 Continued

Name	Age	App used[a]	Relationship status	Sexual identity (applicable to queer informants)	Education	Occupation	Monthly income[b]
Mali	25	Blued, Aloha	Single	Queer	Bachelor's	Graduate student and lawyer	Low
Norman	26	Blued, Aloha	Single	Homosexual/gay	Master's	Education consultant	Midrange
Yuan	26	Blued	Single	Gay/male homosexual	Bachelor's	Education consultant	High
Lasong	26	Blued	Dating	Male *tongzhi*	Bachelor's	Banking executive	Midrange
River	26	Blued	Single	Gay	Bachelor's	Laboratory technician	Midrange
Johnny	27	Blued, Aloha	Dating	*Tongzhi*	Bachelor's	Marketing executive	Midrange
Huajun	28	Blued, Fanbaizi	Dating	Gay	Bachelor's	Personal trainer	Midrange
Jerry	28	Blued, Tinder, OkCupid	Single	Queer	Bachelor's	Computer scientist	Low

Continued

Table A.1 Continued

Name	Age	App used[a]	Relationship status	Sexual identity (applicable to queer informants)	Education	Occupation	Monthly income[b]
Queer women (interviewed in 2018)							
Amy	18	Rela	Dating (with a man)	Bisexual	Secondary school	Telephone debt collector	Low
Alina	19	Rela, Soul	Single	Bisexual	Secondary school	Undergraduate student	Low
Magda	19	Rela	Single	Female homosexual	Secondary school	Undergraduate student	Low
Xiu	23	Rela	Single	Lesbian	Associate's	Unemployed	Low
Shawn	24	Rela, Lesdo	Dating	Queer	Bachelor's	Branding executive	Midrange
Xiaoyan	24	Rela	Single	Lesbian	Bachelor's	Real estate agent	Midrange
Charlie	24	Rela, Lesdo	Dating	Lesbian	Associate's	Social worker	Low
Xi	24	Rela, Lesdo, Lespark, Tantan	Single	Bisexual	Master's	Counselor	Midrange
Dada	25	Rela, Lespark	Dating	Bisexual	Secondary school	Real estate agent	Low

Table A.1 Continued

Name	Age	App used[a]	Relationship status	Sexual identity (applicable to queer informants)	Education	Occupation	Monthly income[b]
Alex	26	Lesdo	Dating	Queer	Bachelor's	Advertising copywriter	Low
Xiaoqing	26	Rela	Single	Female *tongzhi*	Bachelor's	Real estate investor	High
Shushu	26	Rela, Lesdo	Dating	Queer	Master's	Lecturer	Low
Jamie	26	Rela	Dating	Female *tongzhi* / lesbian	Bachelor's	Construction supervisor	Low
Becky	34	Lesdo, Lespark	Dating	Bisexual	Master's	IT maintenance technician	Low

Notes: a. These are the apps that my informants had been using three months prior to my interviews. b. Low: CNY5,000 (~USD725) or below; midrange: CNY5,001 to 10,000 (~USD725 to 1,450); high: CNY10,001 (~USD1,450) or above. c. Wenwei declined to disclose her age. My guess is that she was in her mid-forties. She also declined to disclose her monthly income.

NOTES

CHAPTER 1

1. Nancy is a pseudonym chosen by my informant. I encouraged all of my informants to provide me with a Chinese or English pseudonym. When they did not have any ideas or the name they came up was too similar to their real name, I created one for them.

2. On February 23, 2018, Momo announced that it had bought Tantan for approximately USD600 million.

3. Some apps, such as Blued, require users to submit their national identity card in order to become a live streamer.

4. These apps often update their features. My description in this book is accurate only at the time of writing.

5. *Yue?* can be considered as the Chinese equivalent of "you up?" from Western dating app culture. The difference is that "you up?" is often sent late at night but there is no temporal constraint for sending *yue?*.

6. In this video, which Michael Stephen Kai Sui shared on Weibo, he imitates twelve men from different countries and regions chatting with each other in Chinese. During the conversation, the phrase *yuepao shenqi* is brought up. At the time of writing, the full video is available at https://v.youku.com/v_show/id_XNDEwMjAxNTQ0 .html?spm=a2h0k.11417342.soresults.dtitle and https://www.youtube.com/watch?v =D9dApqW8WZk&list=RDD9dApqW8WZk&start_radio=1&t=0. Alternatively, readers can search for "Mike Sui's 12 Beijingers" on YouTube.

7. Here are a handful of studies that fall under each issue. (1) Why do people use dating apps, and what are their demographic and psychographic characteristics (e.g., L. S. Chan, 2017a, 2017b, 2019; Sumter & Vandenbosch, 2019; Timmermans & De Caluwé, 2017b)? (2) What do people use dating apps for, and what affects their various uses (e.g., Miller, 2015b; Solis & Wong, 2019; Sumter, Vandenbosch, & Ligtenberg, 2017; Timmermans & De Caluwé, 2017a; Van De Wiele & Tong, 2014)? (3) How do people represent themselves on dating apps, and what are the implications

of self-representation and face-to-face encounters (Birnholtz, Fitzpatrick, Handel, & Brubaker, 2014; Blackwell, Birnholtz, & Abbott, 2014; L. S. Chan, 2016; Duguay, 2017; Fitzpatrick, Birnholtz, & Brubaker, 2015; Miller, 2015a)? (4) How do people develop relationships on dating apps (e.g., L. S. Chan, 2018a; De Seta & Zhang, 2015; Fitzpatrick & Birnholtz, 2018; Hobbs, Owen, & Gerber, 2017; Lefebvre, 2018; Licoppe, Rivière, & Morel, 2016; Timmermans & Courtois, 2018; Tong, Hancock, & Slatcher, 2016; Yeo & Fung, 2018; Zhang & Erni, 2018)? (5) What are the risks and concerns associated with the use of dating apps (e.g., Choi, Wong, & Fong, 2017; Corriero & Tong, 2016; Landovitz et al., 2013; Lutz & Ranzini, 2017; Rice et al., 2012; Sawyer, Smith, Benotsch, 2018; Yeo & Ng, 2016)?

8. The idea of intimacy has been explored in multiple disciplines. My understanding of intimacy is based mainly on sociological literature, such as Anthony Giddens's (1992) *The Transformation of Intimacy: Sexuality, Love, and Eroticism in Modern Societies* and Zygmunt Bauman's (2003) *Liquid Love: On the Fragility of Human Bonds*.

9. The phrase "the personal is political" has often been attributed to the American feminist Carol Hanisch. However, Hanisch (2006) explains that the phrase was coined by Shulamith Firestone and Anne Koedt, the editors of *Notes from the Second Year: Women's Liberation*, in which Hanisch's article on consciousness-raising appears.

10. I was inspired by the "Wittgenstein trick" (Becker, 1998, p. 139): what would be left if I were to take away personal relationships from dating apps?

11. I treat network publics and networked individualism, a concept elaborated in Lee Rainie and Barry Wellman (2012), as two sides of the same coin. Rainie and Wellman argue that in the digital era we are detached from traditional communities but are reembedded in several relational networks. Their focus is on the constitution of selfhood in network society. The concept of network publics, alternatively, highlights not the self but the collective that the self belongs to.

12. Other studies of dating apps falling under the critical tradition have explored issues such as data cultures (Albury, Burgess, Light, Race, & Wilken, 2017), the creation of queer communities (Baudinette, 2019; Crooks, 2013; Roth, 2014), the politics of play and games (Race, 2015; Wang, 2020), gender politics (Bivens & Hoque, 2018), and state censorship (T. Liu, 2016).

13. Further, the same environmental feature can theoretically provide different affordances for different animals. For instance, a water surface affords support only to water bugs, not to human beings (Gibson, 1979).

14. Donald Norman (1988) popularized the contrast between affordances and constraints. To him, affordances are positive and physical, whereas constraints are negative and, in addition to physical, cultural. An example he gives to illustrate the contrast uses Legos to build a police motorcycle. The physical shape of a Lego piece both affords and constrains how we can fit one piece onto another piece; however, our cultural understanding that "yellow is a headlight" further restricts how we can build the motorcycle.

15. In my original formulation, I named this affordance *visual dominance*. However, visual dominance is better conceptualized as an outcome resulting from visibility (see Evans, Pearce, Vitak, & Treem, 2017).

16. For an overview of the global transformation of intimacy, see Mark Padilla, Jennifer Hirsch, Miguel Munoz-Laboy, Robert Sember, and Richard Parker (2007).

17. Cantonese and Putonghua are so different that they are mutually unintelligible; therefore, Cantonese has become a site for identity politics. In July 2010, a

Guangzhou government officer suggested a major local television station start using Putonghua instead of Cantonese in its news program. More than a thousand people, mostly in their twenties and thirties, gathered to protest against the proposal. The protest ended peacefully, but online discussion of the protest was censored afterward (E. Wong, 2010).

18. I count Nancy as a straight woman because she identified herself as straight when I interviewed her.

19. One of my heterosexual female informants, Chloe, had just stopped using dating apps before I conducted the interview with her.

CHAPTER 2

1. In *xiangqin* 相亲 (matching), parents take the lead to preselect several candidates for their adult children. The parents then arrange for their children to meet with these candidates one by one over a meal. During this meeting, the young adults introduce themselves to each other, and if they like each other, they will begin to date. One way for parents to identify suitable candidates is through their personal networks. Another way is to attend the matching activities in the parks.

2. Regarding the use of dating apps, the dominant theoretical framework that recent dating app studies have been relying on is uses and gratifications theory (Katz, Blumler, & Gurevitch, 1973). I provide my critique to this approach in the concluding chapter.

3. In rural China, it is normal for a couple to live with the husband's parents, a practice called *patrilocality*. In some of the rural-to-urban couples interviewed by Susanne Yuk-Ping Choi and Yinni Peng (2016), the wives negotiated to stay closer to their natal families. However, some men split up with their girlfriends because their girlfriends refused to live with their parents.

4. China's census data from 2000 is available at http://www.stats.gov.cn/tjsj/pcsj/rkpc/5rp/index.htm. Its 2010 census data is available at http://www.stats.gov.cn/tjsj/pcsj/rkpc/6rp/indexch.htm.

5. Yifei Shen (2011) argues that while fathers are losing power to their daughters, husbands are still in a more dominant position than their wives. She calls this *postpatriarchy*.

6. The algorithmic alert system has its limitations. It does not detect semantic nuances. For example, an alert was generated when I typed *yue ge fangwen* 约个访问 (arrange an interview). Meanwhile, when I broke down the character *yue* 约 into two characters 糸 and 勺, the system did not recognize the meaning.

7. Readers who start reading from this chapter may want to look at the introductory section of chapter 1 to familiarize themselves with the various features that dating apps in China offer that are not common in Western apps.

8. Very often, people say or do things in online interactions that they would not say or do in face-to-face interactions. John Suler (2004) calls this phenomenon the online disinhibition effect. Dissociative anonymity, invisibility, asynchronicity, solipsistic introjection, dissociative imagination, and minimization of authority contribute to this effect.

9. Gayle Rubin (1984) contends that, due to the influence of Christianity, Western culture "always treats sex with suspicion. It construes and judges almost any sexual practice in terms of its worst possible expression. Sex is presumed guilty until proven innocent" (p. 278).

CHAPTER 3

1. One of the sex-related jokes Victor shared with me goes as follows: "A man asks a woman to pay for a pack of cigarettes. The woman says, 'Tonight, one minute is worth one dollar. How expensive the cigarettes you want to smoke depends on your performance.' The man later is found to be smoking a pack of cigarettes that costs two dollars."

2. Comprehensive critiques of hegemonic masculinity can be found in Øystein Gullvåg Holter (2003) and Richard Howson (2006).

3. According to Raewyn Connell and James Messerschmidt (2005), hegemonic masculinities may manifest at the local, regional, and global levels, as follows: "1) Local: constructed in the arenas of face-to-face interaction of families, organizations, and immediate communities, as typically found in ethnographic and life-history research; 2) Regional: constructed at the level of the culture or the nation-state, as typically found in discursive, political, and demographic research; and 3) Global: constructed in transnational arenas such as world politics and transnational business and media, as studied in the emerging research on masculinities and globalization" (p. 849).

4. *Cha chaan teng* is a Hong Kong–style fast food that is very popular in Guangzhou. Victor mentions *cha chaan teng* and coffee shops. Both are fast food.

5. Including prostitution in corporate entertainment is not uncommon in East Asia (Norma, 2011; Zhou, 2006).

6. Media psychologists are interested in how choice overload affects human decisions. Jonathan D'Angelo and Catalina Toma (2017) experimentally assessed the effect of the number of choices on the level of satisfaction with one's partner selection in online dating. The participants in one group selected a date prospect from a pool of six, and another group chose from a pool of twenty-four. After one week, the participants in the large-choice condition were less satisfied with their selection and more likely to reverse their selection than those in the small-choice condition. The experiment demonstrates that more choices do not always mean better results.

7. As I point out in chapter 2, on dating apps, objectification goes both ways. Some of my young female informants exercised their female gaze, too.

CHAPTER 4

1. In this chapter, I use *queer male informants* as an umbrella term to refer to all of my nonheterosexual-identifying male informants. Whenever I refer to individual informants, I follow how they wanted to be referred to. *Tongzhi*, which literally means "comrades" in Communist parlance, is a gender-neutral label used by queer people in China to refer to themselves. My informants also used Western identity labels, such as *ku'er* 酷儿 (queer). The English word *gay* is common enough to be used as an everyday sexual identifier, but it is also translated as *ji* 基. Accordingly, gay men describe themselves as gay, *ji*, or *jilao* 基佬 (*lao* is a Chinese colloquial term for "guy"). Hongwei Bao (2018) recently found that people from different social and economic backgrounds preferred different labels. For instance, people who had a more cosmopolitan outlook addressed themselves as gay or *ji*, whereas those who worked for LGBTQ-serving organizations preferred *tongzhi*. I did not see such a pattern among my informants.

2. Only one queer female informant and no straight informants told me that they had repetitively deleted and installed a dating app for reasons unrelated to finding a romantic partner.

3. For example, *duanxiu* 断袖 (cutting a sleeve) and *fentao* 分桃 (sharing a peach) have appeared in Chinese classics. The former refers to Mizi Xia, a male concubine who shared a piece of peach with the Duke Ling of Wey, who ruled the state of Wey from 534 BC to 492 BC. The latter refers to Emperor Ai of the Han empire from 534 BC to 492 BC, who cut off his sleeve to avoid waking up his male partner who had been sleeping on it. These terms do not carry negative connotations. For a historical view of Chinese male homosexuality, see Bret Hinsch (1990).

4. There was also liberal discourse about homosexuality among some of the translated materials. According to Tze-Ian Sang (2003), "It would be more accurate to say that, in the 1920s, the literate, urban Chinese public had no definite opinion on the nature of same-sex love and that the Chinese intellectuals who had access to theories of homosexuality written in other languages had a great deal of room in which to maneuver" (p. 106).

5. Unlike *duanxiu* and *fentao*, *pi* and *renyao* carried a stigma. In particular, *renyao* was used as an umbrella term to describe sexually deviant men, including "sexually prematurely developed boy[s], cross dressers, intersex people who might have sex with both men and women, Peking opera *dan* actors, male prostitutes, and any men who behaved and dressed in a feminine fashion and had sex with other men" (Kang, 2009, p. 34).

6. Due to the rarity of physical homophobic expression in China, Wah-Shan Chou (2000) argues that there is a cultural tolerance of homosexuality in China. However, this celebratory tone has been heavily criticized by other scholars, who regard tolerance as a silencing tool (Kam, 2013; Liu & Ding, 2005).

7. Geng Le also started danlan.org, one of the earliest gay online forums, in 2000.

8. Zank was shut down in 2017 after the government cracked down on the pornographic content in its live-streaming.

9. Eve Kosofsky Sedgwick's interpretation of Silvan Tomkins can be found in *Shame and Its Sisters: A Silvan Tomkins Reader* (Sedgwick & Frank, 1995).

10. Accordingly, shame is the opposite of pridefulness, whereas guilt is the opposite of pride.

11. For example, Silvan Tomkins (1962, 1963) proposed nine building blocks of affects that can form more complex emotions. These include enjoyment-joy, interest-excitement, fear-terror, distress-anguish, anger-rage, shame-humiliation, disgust, dissmell, and surprise-startle. Robert Plutchik (1979) suggested that there are eight, not nine, primary emotions. These are joy, sadness, acceptance, disgust, fear, anger, surprise, and anticipation. Some of the emotions identified by Plutchik were also identified by Tomkins; some were similar but not completely the same, like sadness in Plutchik's model and distress in Tomkins's.

12. Western media have criticized gay dating apps such as Grindr for publishing the exact distances between users because any user can be tracked down to their exact position through triangulation ("It Is Still Possible," 2018).

13. When I logged onto Blued on July 21, 2018, in Guangzhou, only 15 of the closest 150 people I could see on my app had verified photographs, reflecting this feature's lack of popularity.

14. Jack'd, owned by the American company Online Buddies at the time of writing, was once popular in China. However, my informants told me that because the app was in English and the connection was unstable, they had gradually moved to locally developed apps.

15. This profile was collected as part of my earlier study (L. S. Chan, 2016) in which I compared 204 Jack'd profiles from the United States and 204 Jack'd profiles from China using geographically stratified sampling.

16. *Dashu* 大叔 (uncle) in Chinese gay culture is equivalent to a daddy in Western gay culture. Both refer to an older man sexually interested in younger men.

17. On August 13, 2020, when I was finalizing this manuscript, the organizers of Shanghai Pride announced indefinite cancellation of all future activities. I could not recontact my informants to get their views on this; however, I agree with Jose Muñoz (2009) that we must keep our hope amid difficult times.

18. I discuss the communal aspect of dating apps in detail in chapter 5.

19. Alan Chan and Sor-Hoon Tan (2004) provide a wonderful overview of filial piety in Chinese thought and history.

CHAPTER 5

1. In April 2019, Tantan, the app for straight people, was also taken down by the government because of its pornographic content. The app was relaunched in June.

2. In this chapter, I use the terms *queer women* and *queer female informants* as umbrella terms to refer to all of my nonheterosexual-identifying female informants. When I refer to individual informants, I follow how they would like to be referred to. Apart from *ku'er* 酷儿 (queer) and *nü tongxinglian zhe* 女同性恋者 (female homosexual), *lala* 拉拉 and *les* (both derived from *lesbians*) are common sexual identities used by queer women in China. Bisexual people often call themselves *bi* or *shuang* 双.

3. All Chinese movies having LGBTQ content must, by definition, be independently produced because the state has banned such content in mainstream movies.

4. According to the Chinese Center for Disease Control and Prevention, women have accounted for only around a quarter of new HIV infection cases every year since 2012 ("2018 nian Zhongguo aizibing ganran renshu," 2018).

5. The analysis of this subsection is informed by the walkthrough method developed by Ben Light, Jean Burgess, and Stefanie Duguay (2018) in which apps are treated as cultural texts. This method "is a way of engaging directly with an app's interface to examine its technological mechanisms and embedded cultural references to understand how it guides users and shapes their experiences" (p. 882). The complete methodology involves analyses of the app's vision, operation model, and governance, and a step-by-step observation of the registration, everyday use, and disconnection of the app.

6. Although media activism is a dominant form of activism in contemporary China (Tan, 2017; G. Yang, 2017), there have been cases of sporadic offline LGBTQ protests. For instance, on August 25, 2009, in People's Park in Guangzhou, which is known to the local gay community as a cruising spot, five policemen ordered gay men to leave the premises. The order from the police force resulted in a confrontation between around a hundred gay men and the five policemen. The policemen eventually left the park. This victory to claim the public space was widely celebrated by regional and international LGBTQ groups (Bao, 2018).

7. Officially, Blued positions itself as a health education app, not a hookup app (Miao & Chan, 2020).

8. An historical discussion of butch-femme differences in American lesbian communities can be found in Elizabeth Lapovsky Kennedy and Madeline Davis's *Boots*

of Leather, Slippers of Gold: The History of a Lesbian Community (1993). In Hong Kong, gender labels of *tb* (tomboy) and *tbg* are used. The shorthand *tbg* represents lesbians who act more feminine and prefer dating a *tb*. Physical appearances often determine one's choice of labels (D. T.-S. Tang, 2015). The labels *T* and *P* came from Taiwan and were introduced to mainland China in the early 2000s (Engebretsen, 2014).

9. These social app videos have three characteristics: they are produced by the apps, they are expected to be watched via the apps, and the use of the apps appears in the main storyline. Some titles produced by Lesdo are *Xiaxue & Weian: Always Miss You* 夏雪薇安之念念不忘 (2016, 34 minutes) and *True Love, Wrong Gender* 错了性别不错爱 (2016, 60 minutes). Rela produced *X-Love* 爱的未知数 (2016, 19 minutes) and *Girls Who Talk to Flowers* 如果花会说话 (2016, 12 minutes) (see Tan, 2018).

10. At the time of writing, all four episodes of *The L Bang* were available on YouTube.

11. James Gibson (1979) differentiated positive and negative affordances that result from the same interaction between the features of the environment and the actor. Accordingly, connecting with the community can be viewed as a positive affordance, whereas complying with communal norms can be viewed as a negative affordance.

12. That said, I am not suggesting that technology is a neutral tool. As I point out in chapter 1, technological artifacts are political (Winner, 1980). The notion of affordance allows us to avoid perceiving the effects of technology as linear and deterministic.

CHAPTER 6

1. In "Sex in Public," Lauren Berlant and Michael Warner (1998) argue that a queer world is different from a community or a group because "it necessarily includes more people than can be identified, more spaces than can be mapped beyond a few reference points, modes of feeling that can be learned rather than experienced as a birthright" (p. 558).

2. I want to highlight that Kangqi was in an open relationship and Damon was single during the time of our interviews. Both were open to having casual sex. Therefore, their use of the term *mudixing* meant not simply someone seeking sex on dating apps but, more specifically, the singularity of such purpose.

3. Chad Van De Wiele and Stephanie Tom Tong (2014) published the first uses and gratifications study on Grindr based on data from worldwide users. Similarly, Brandon Miller (2015b) conducted a worldwide study on men who have sex with men. Elisabeth Timmermans and Elien De Caluwé (2017a) and Sindy Sumter, Laura Vandenbosch, and Loes Ligtenberg (2017) explore the motives for using Tinder based on American/Belgian samples and a Dutch sample, respectively. Randy Jay Solis and Ka Yee Wong (2019) look at dating apps in China.

4. Charles Ragin and Peer Fiss (2017) also argue that conventional regression-based interaction analysis is less compatible with intersectionality theory. See their work for an example of taking an alternative mathematical approach, qualitative comparative analysis, to examine how combinations of factors lead to poverty.

5. Alipay is a mobile wallet owned by the Alibaba Group.

REFERENCES

Abidin, C. (2016). Agentic cute (^.^): Pastiching East Asian cute in influencer commerce. *East Asian Journal of Popular Culture, 2*(1), 33–47. https://doi.org/10.1386/eapc.2.1.33_1

Ahmed, S. (2004a). Collective feelings: Or, the impressions left by others. *Theory, Culture and Society, 21*(2), 25–42. https://doi.org/10.1177/0263276404042133

Ahmed, S. (2004b). *The cultural politics of emotion.* Edinburgh, UK: Edinburgh University Press.

Ahmed, S. (2010). *The promise of happiness.* Durham, NC: Duke University Press.

Albury, K., Burgess, J., Light, B., Race, K., & Wilken, R. (2017). Data cultures of mobile dating and hook-up apps: Emerging issues for critical social science research. *Big Data and Society, 4*(2), 1–11. https://doi.org/10.1177/2053951717720950

Albury, K., & Byron, P. (2016). Safe on my phone? Same-sex attracted young people's negotiations of intimacy, visibility, and risk on digital hook-up apps. *Social Media and Society, 2*(4), 1–10. https://doi.org/10.1177/2056305116672887

Analysys. (2016). *2016 Zhongguo tongzhi shejiao yingyong shichang zhuanti yanjiu baogao 2016 中国同志社交应用市场专题研究报告* [2016 Chinese *tongzhi* social app market research report]. Retrieved from https://www.useit.com.cn/thread-11431-1-1.html

Anderson, W. (2012). Asia as method in science and technology studies. *East Asian Science, Technology and Society, 6*(4), 445–451. https://doi.org/10.1215/18752160-1572849

Armstrong, H. L., & Reissing, E. D. (2015). Women's motivations to have sex in casual and committed relationships with male and female partners. *Archives of Sexual Behavior, 44*(4), 921–934. https://doi.org/10.1007/s10508-014-0462-4

Attwood, F., Hakim, J., & Winch, A. (2017). Mediated intimacies: Bodies, technologies and relationships. *Journal of Gender Studies, 26*(3), 249–253. https://doi.org/10.1080/09589236.2017.1297888

Ayres, T. (1999). China doll: The experience of being a gay Chinese Australian. *Journal of Homosexuality, 36*(3–4), 87–97. https://doi.org/10.1300/J082v36n03_05

Bailey, B. (1988). *From front porch to back seat: Courtship in twentieth-century America.* Baltimore, MD: Johns Hopkins University Press.

Banet-Weiser, S. (2018). *Empowered: Popular feminism and popular misogyny.* Durham, NC: Duke University Press.

Bao, H. (2018). *Queer comrades: Gay identity and tongzhi activism in postsocialist China.* Copenhagen, Denmark: Nordic Institute of Asian Studies.

Baudinette, T. (2019). Gay dating applications and the production/reinforcement of queer space in Tokyo. *Continuum: Journal of Media and Cultural Studies, 33*(1), 93–104. https://doi.org/10.1080/10304312.2018.1539467

Bauman, Z. (2003). *Liquid love: On the frailty of human bonds.* Cambridge, UK: Polity.

Baym, N. (2010). *Personal connection in the digital age.* Cambridge, UK: Polity.

Becker, H. S. (1998). *Tricks of the trade: How to think about your research while you're doing it.* Chicago, IL: University of Chicago Press.

Belton, P. (2018, February 13). What have dating apps really done for us? *BBC News.* Retrieved from https://www.bbc.com/news/business-42988025

Berlant, L., & Warner, M. (1998). Sex in public. *Critical Inquiry, 24*(2), 547–566. https://doi.org/10.1086/448884

Berry, C., & Martin, F. (2003). Syncretism and synchronicity: Queer'n'Asian cyberspace in 1990s Taiwan and Korea. In C. Berry, F. Martin, & A. Yue (Eds.), *Mobile cultures: New media in queer Asia* (pp. 87–114). Durham, NC: Duke University Press.

Berry, C., Martin, F., & Yue, A. (Eds.). (2003). *Mobile cultures: New media in queer Asia.* Durham, NC: Duke University Press.

Bertaux, D. (1981). From the life-history approach to the transformation of sociological practice. In D. Bertaux (Ed.), *Biography and society: The life history approaches in the social sciences* (pp. 29–45). Beverley Hill, CA: Sage.

Bhandari, B. (2017, May 20). At wedding market, mothers of gays and lesbians face resistance. *Sixth Tone.* Retrieved from http://www.sixthtone.com/news/1000233/at-wedding-market%2C-mothers-of-gays-and-lesbians-face-resistance

Bird, S. R. (1996). Welcome to the men's club: Homosociality and the maintenance of hegemonic masculinity. *Gender and Society, 10*(2), 120–132. https://doi.org/10.1177/089124396010002002

Birnholtz, J. (2010). Adopt, adapt, abandon: Understanding why some young adults start, and then stop, using instant messaging. *Computers in Human Behavior, 26*(6), 1427–1433. https://doi.org/10.1016/j.chb.2010.04.021

Birnholtz, J., Fitzpatrick, C., Handel, M., & Brubaker, J. R. (2014). Identity, identification and identifiability: The language of self-presentation on a location-based mobile dating app. *Proceedings of the 16th International Conference on Human-Computer Interaction with Mobile Devices and Services* (pp. 3–12). https://doi.org/10.1145/2628363.2628406

Bivens, R., & Hoque, A. S. (2018). Programming sex, gender, and sexuality: Infrastructural failures in the "feminist" dating app Bumble. *Canadian Journal of Communication, 43*(3), 441–459. https://doi.org/10.22230/cjc.2018v43n3a3375

Blackshaw, T. (2010). *Key concepts in community studies.* London, UK: Sage.

Blackwell, C., Birnholtz, J., & Abbott, C. (2014). Seeing and being seen: Co-situation and impression formation using Grindr, a location-aware gay dating app. *New Media and Society, 17*(7), 1117–1136. https://doi.org/10.1177/1461444814521595

Boellstorff, T., Nardi, B., Pearce, C., & Taylor, T. L. (2012). *Ethnography and virtual worlds: A handbook of method.* Princeton, NJ: Princeton University Press.

Bourdieu, P., & Wacquant, L. J. (1992). *An invitation to reflexive sociology.* Chicago, IL: University of California Press.

boyd, d. (2011). Social network sites as networked publics: Affordances, dynamics, and implications. In Z. Papacharissi (Ed.), *Networked self: Identity, community, and culture on social network sites* (pp. 39–58). New York, NY: Routledge.

Brait, E. (2015, September 29). Tinder and Grindr outraged over STD testing billboards that reference apps. *The Guardian.* Retrieved from https://www.theguardian.com/technology/2015/sep/29/tinder-grindr-std-testing-aids-healthcare-foundation-billboards

Bridges, T. (2014). A very "gay" straight? Hybrid masculinities, sexual aesthetics, and the changing relationship between masculinity and homophobia. *Gender and Society, 28*(1), 58–82. https://doi.org/10.1177/0891243213503901

Brubaker, J. R., Annany, M., & Crawford, K. (2016). Departing glances: A sociotechnical account of "leaving" Grindr. *New Media and Society, 18*(3), 373–390. https://doi.org/10.1177/1461444814542311

Burgess, J., Cassidy, E., Duguay, S., & Light, B. (2016). Making digital cultures of gender and sexuality with social media. *Social Media+ Society, 2*(4), 1–4. https://doi.org/10.1177/2056305116672487

Butler, J. (1999). *Gender trouble: Feminism and the subversion of identity.* New York, NY: Routledge.

Butler, J. (2004). *Undoing gender.* New York, NY: Routledge.

Cabañes, J. V. A., & Collantes, C. F. (2020). Dating apps as digital flyovers: Mobile media and global intimacies in a postcolonial city. In J. V. A. Cabañes & C. S. Uy-Tioco (Eds.), *Mobile media and social intimacies in Asia: Reconfiguring local ties and enacting global relationships* (pp. 97–114). Dordrecht, the Netherlands: Springer.

Cabañes, J. V. A., & Uy-Tioco, C. S. (2020). *Mobile media and social intimacies in Asia: Reconfiguring local ties and enacting global relationships.* Dordrecht, the Netherlands: Springer.

Campbell, J. E. (2004). *Getting it on online: Cyberspace, gay male sexuality, and embodied identity.* New York, NY: Harrington Park Press.

Carey, J. (1967). Harold Adams Innis and Marshall McLuhan. *The Antioch Review, 27*(1), 5–39. https://doi.org/10.2307/4610816

Cassidy, E. (2018). *Gay men, identity, and social media: A culture of participatory reluctance.* New York, NY: Routledge.

Cavalcante, A. (2019). Tumbling into queer utopias and vortexes: Experiences of LGBTQ social media users on Tumblr. *Journal of Homosexuality, 66*(12), 1715–1735. https://doi.org/10.1080/00918369.2018.1511131

Chan, A. K.-L., & Tan, S.-H. (2004). *Filial piety in Chinese thought and history.* London, UK: Routledge.

Chan, L. S. (2016). How sociocultural context matters in self-presentation: A comparison of US and Chinese profiles on Jack'd, a mobile dating app for men who have sex with men. *International Journal of Communication, 10*, 6040–6059.

Chan, L. S. (2017a). The role of gay identity confusion and outness in sex-seeking on mobile dating apps among men who have sex with men: A conditional process analysis. *Journal of Homosexuality, 64*(5), 622–637. https://doi.org/10.1080/00918369.2016.1196990

Chan, L. S. (2017b). Who uses dating apps? Exploring the relationships among trust, sensation-seeking, smartphone use, and the intent to use dating apps based on the integrative model. *Computers in Human Behavior, 72*, 246–258. https://doi.org/10.1016/j.chb.2017.02.053

Chan, L. S. (2018a). Ambivalence in networked intimacy: Observations from gay men using mobile dating apps. *New Media and Society, 20*(7), 2566–2581. https://doi.org/10.1177/1461444817727156

Chan, L. S. (2018b). Liberating or disciplining? A technofeminist analysis of the use of dating apps among women in urban China. *Communication, Culture and Critique, 11*(2), 298–314. https://doi.org/10.1093/ccc/tcy004/4956846

Chan, L. S. (2019). Paradoxical associations of masculine ideology and casual sex among heterosexual male geosocial networking app users in China. *Sex Roles, 81,* 456–499. https://doi.org/10.1007/s11199-019-1002-4

Chan, L. S. (2020). Multiple uses and anti-purposefulness on Momo, a Chinese dating/social app. *Information, Communication and Society, 23*(10), 1515–1530. https://doi.org/10.1080/1369118X.2019.1586977

Chen, G. M. (2011). Tweet this: A uses and gratifications perspective on how active Twitter use gratifies a need to connect with others. *Computers in Human Behavior, 27*(2), 755–762. https://doi.org/10.1016/j.chb.2010.10.023

Cheung, C.-F. (1999). Western love, Chinese *qing*: A philosophical interpretation of the idea of love in *Romeo and Juliet* and *The Butterfly Lover*. *Journal of Chinese Philosophy, 26*(4), 469–488. https://doi.org/10.1111/j.1540-6253.1999.tb00553.x

China Internet Network Information Center. (2017). *Statistical Report on Internet Development in China (January 2017).* Retrieved from https://cnnic.com.cn/IDR/ReportDownloads/201706/P020170608523740585924.pdf

Chinese gay dating app Blued halts registration after underage HIV report. (2019, January 9). *South China Morning Post.* Retrieved from https://www.scmp.com/tech/apps-social/article/2180895/chinese-gay-dating-app-blued-halts-registration-after-underage-hiv

Chinn, S. E. (2012). Queer feelings/feeling queer: A conversation with Heather Love about politics, teaching, and the "dark, render thrills" of affect. *Transformations: The Journal of Inclusive Scholarship and Pedagogy, 22*(2), 124–131. https//doi.org/10.17613/M6M53K

Choi, E. P. H., Wong, J. Y. H., & Fong, D. Y. T. (2017). The use of social networking applications of smartphone and associated sexual risks in lesbian, gay, bisexual, and transgender populations: A systematic review. *AIDS Care, 29*(2), 145–155. https://doi.org/10.1080/09540121.2016.1211606

Choi, G., & Chung, H. (2013). Applying the technology acceptance model to social networking sites (SNS): Impact of subjective norm and social capital on the acceptance of SNS. *International Journal of Human–Computer Interaction, 29*(10), 619–628. https://doi.org/10.1080/10447318.2012.756333

Choi, S. Y.-P., & Luo, M. (2016). Performative family: Homosexuality, marriage and intergenerational dynamics in China. *British Journal of Sociology, 67*(2), 206–208. https://doi.org/10.1111/1468-4446.12196

Choi, S. Y.-P., & Peng, Y. (2016). *Masculine compromise: Migration, family, and gender in China.* Berkeley, CA: University of California Press.

Chong, E. S., Zhang, Y., Mak, W. W., & Pang, I. H. (2015). Social media as social capital of LGB individuals in Hong Kong: Its relations with group membership, stigma, and mental well-being. *American Journal of Community Psychology, 55*(1–2), 228–238. https://doi.org/10.1007/s10464-014-9699-2

Chou, W.-S. (2000). *Tongzhi: Politics of same-sex eroticism in Chinese societies.* New York, NY: Haworth Press.

Choy, C. H. Y. (2018). Smartphone apps as cosituated closets: A lesbian app, public/ private spaces, mobile intimacy, and collapsing contexts. *Mobile Media and Communication, 6*(1), 88–107. https://doi.org/10.1177/2050157917727803

Cipolla, C., Gupta, K., Rubin, D. A., & Willey, A. (Eds.). (2017). *Queer feminist science studies: A reader.* Seattle, WA: University of Washington Press.

Cohn, C. (1987). Sex and death in the rational world of defense intellectuals. *Signs: Journal of Women in Culture and Society, 12*(4), 687–718. https://doi.org/10.1086/494362

Connell, R. W. (1987). *Gender and power: Society, the person and sexual politics.* Stanford, CA: Stanford University Press.

Connell, R. W., & Messerschmidt, J. W. (2005). Hegemonic masculinity: Rethinking the concept. *Gender and Society, 19*(6), 829–859. https://doi.org/10.1177/0891243205278639

Connell, R. W., & Wood, J. (2005). Globalization and business masculinities. *Men and Masculinities, 7*(4), 347–364. https://doi.org/10.1177/1097184X03260969

Conte, M. (2017). *More fats, more femmes, and no whites: A critical examination of fatphobia, femmephobia and racism on Grindr.* Unpublished master's thesis, Carleton University. Retrieved from https://doi.org/10.22215/etd/2017-12122

Cooper, M. (2000). Being the "go-to guy": Fatherhood, masculinity, and the organization of work in Silicon Valley. *Qualitative Sociology, 23*(4), 379–405. https://doi.org/10.1023/A:1005522707921

Corriero, E. F., & Tong, S. T. (2016). Managing uncertainty in mobile dating applications: Goals, concerns of use, and information seeking in Grindr. *Mobile Media and Communication, 4*(1), 121–141. https://doi.org/10.1177/2050157915614872

Crenshaw, K. (1989). Demarginalizing the intersection of race and sex: A black feminist critique of antidiscrimination doctrine, feminist theory and antiracist politics. *University of Chicago Legal Forum, 1989*(1), 139–167.

Crooks, R. N. (2013). The rainbow flag and the green carnation: Grindr in the gay village. *First Monday, 18*(11). https://doi.org/10.5210/fm.v18i11.4958

Curry, T. J. (1993). A little pain never hurt anyone: Athletic career socialization and the normalization of sports injury. *Symbolic Interaction, 16*(3), 273–290. https://doi.org/10.1525/si.1993.16.3.273

Cvetkovich, A., & Kellner, D. (1997). Introduction: Thinking global and local. In A. Cvetkovich & D. Kellner (Eds.), *Articulating the global and the local: Globalization and cultural studies* (pp. 1–30). Boulder, CO: Westview Press.

D'Angelo, J. D., & Toma, C. L. (2017). There are plenty of fish in the sea: The effects of choice overload and reversibility on online daters' satisfaction with selected partners. *Media Psychology, 20*(1), 1–27. https://doi.org/10.1080/15213269.2015.1121827

Dating app Once introduces Black Mirror–style feature for rating men and empowering women. (2018, March 1). *SF Weekly.* Retrieved from http://www.sfweekly.com/sponsored/dating-app-once-introduces-black-mirror-style-feature-for-rating-men-and-empowering-women/#

Davis, D. (2014). On the limits of personal autonomy. In D. Davis & S. Friedman (Eds.), *Wives, husbands, and lovers: Marriage and sexuality in Hong Kong, Taiwan, and urban China* (pp. 41–61). Stanford, CA: Stanford University Press.

Davis, D., & Friedman, S. (2014). Deinstitutionalizing marriage and sexuality. In D. Davis & S. Friedman (Eds.), *Wives, husbands, and lovers: Marriage and sexuality in Hong Kong, Taiwan, and urban China* (pp. 1–38). Stanford, CA: Stanford University Press.

Davis, F. (1989). Perceived usefulness, perceived ease of use, and user acceptance of information technology. *MIS Quarterly, 13*(3), 319–340. https://doi.org/10.2307/249008

Deng, I. (2018, June 13). Live-streaming video helped this Chinese hook-up app surpass US$10 billion in market value. *South China Morning Post.* Retrieved from https://www.scmp.com/tech/enterprises/article/2150473/live-streaming-video-helped-chinese-hook-app-surpass-us10-billion

De Seta, G., & Zhang, G. (2015). Stranger stranger or lonely lonely? Young Chinese and dating apps between the locational, the mobile and the social. In I. A. Degim, J. Johnson, & T. Fu (Eds.), *Online courtship: Interpersonal interactions across borders* (pp. 167–185). Amsterdam, The Netherlands: Institute of Network Cultures.

Dobson, A. S., Robards, B., & Carah, N. (2018). Digital intimate publics and social media: Towards theorizing public lives on private platforms. In A. S. Dobson, B. Robards, & N. Carah (Eds.), *Digital intimate publics and social media* (pp. 3–27). Cham, Switzerland: Palgrave Macmillan.

D'Onofrio, J. (2018, December 3). A better, more positive Tumblr [Web log post]. Retrieved from https://staff.tumblr.com/post/180758987165/a-better-more-positive-tumblr

Dou, E. (2015, November 9). As attitudes in China begin to shift, gay dating app Blued sees green. *The Wall Street Journal.* Retrieved from http://blogs.wsj.com/chinarealtime/2015/11/09/as-attitudes-in-china-begin-to-shift-gay-dating-app-blued-sees-green/

Drell, C. (2017, April 19). Five years later, what have dating apps really done for us? *Glamour.* Retrieved from https://www.glamour.com/story/what-have-dating-apps-really-done-for-us

Duggan, L. (2003). *The twilight of equality? Neoliberalism, cultural politics, and the attack on democracy.* Boston, MA: Beacon.

Duguay, S. (2017). Dressing up Tinderella: Interrogating authenticity claims on the mobile dating app Tinder. *Information, Communication and Society, 20*(3), 351–367. https://doi.org/10.1080/1369118X.2016.1168471

Duguay, S. (2019). "There's no one new around you": Queer women's experiences of scarcity in geospatial partner-seeking on Tinder. In C. J. Nash & A. Gorman-Murray (Eds.), *The geographies of digital sexuality* (pp. 93–114). Singapore: Palgrave Macmillan.

Duportail, J. (2017, September 26). I asked Tinder for my data. It sent me 800 pages of my deepest, darkest secrets. *The Guardian.* Retrieved from https://www.theguardian.com/technology/2017/sep/26/tinder-personal-data-dating-app-messages-hacked-sold

Dutot, V. (2014). Adoption of social media using technology acceptance model: The generational effect. *International Journal of Technology and Human Interaction, 10*(4), 18–35. https://doi.org/10.4018/ijthi.2014100102

Edmunds, S. (2017, September 7). Blued turns to live-streaming after hitting $15m in revenue last year. *Global Dating Insights.* Retrieved from https://globaldatinginsights.com/2017/09/07/blued-turns-to-live-streaming-after-hitting-15m-revenue-last-year/

Eklund, L. (2018). Filial daughter? Filial son? How China's young urban elite negotiate intergenerational obligations. *NORA - Nordic Journal of Feminist and Gender Research, 26*(4), 295–312. https://doi.org/10.1080/08038740.2018.1534887

Ellison, N. B., Hancock, J. T., & Toma, C. L. (2012). Profile as promise: A framework for conceptualizing veracity in online dating self-presentations. *New Media and Society, 14*(1), 45–62. https://doi.org/10.1177/1461444811410395

Elster, J. (1999). *Strong feelings: Emotion, addiction, and human behavior.* Cambridge, MA: MIT Press.

Engebretsen, E. (2014). *Queer women in urban China: An ethnography.* New York, NY: Routledge.

Evans, H. (2008). Sexed bodies, sexualized identities, and the limits of gender. *China Information, 22*(2), 361–386. https://doi.org/10.1177/0920203X08091550

Evans, S., Pearce, K., Vitak, J., & Treem, J. (2017). Explicating affordances: A conceptual framework for understanding affordances in communication research. *Journal of Computer-Mediated Communication, 22*(1), 35–52. https://doi.org/10.1111/jcc4.12180

Fan, C. C. (2003). Rural-urban migration and gender division of labor in transitional China. *International Journal of Urban and Regional Research, 27*(1), 24–47. https://doi.org/10.1111/1468-2427.00429

Fan, P. (2015). Challenging authorities and building community culture: Independent queer film making in China and the China Queer Film Festival Tour,

2008–2012. In E. L. Engebretsen & W. F. Schroeder (Eds.), *Queer/Tongzhi China: New perspectives on research, activism and media cultures* (pp. 81–88). Copenhagen, Denmark: NIAS Press.

Farrer, J. (2002). *Opening up: Youth sex culture and market reform in Shanghai.* Chicago, IL: University of Chicago Press.

Faulkner, W. (2001). The technology question in feminism: A view from feminist technology studies. *Women's Studies International Forum, 24*(1), 79–95. https://doi.org/10.1016/S0277-5395(00)00166-7

Fei, H.-T. (1939). *Peasant life in China: A field study of country life in the Yangtze valley.* New York, NY: E. P. Dutton.

Feldshuh, H. (2018). Gender, media, and myth-making: Constructing China's leftover women. *Asian Journal of Communication, 28*(1), 38–54. https://doi.org/10.1080/01292986.2017.1339721

Feng, J. (2018, December 6). Guangzhou Gender and Sexuality Education Center shuts down. *SupChina.* Retrieved from https://supchina.com/2018/12/06/guangzhou-gender-and-sexuality-education-center-shuts-down/?fbclid=IwAR1zpKAg2fKelfZHwSImQdragfw7tnnzqWj_bKat7SyPTq0wWLHIGGY9_nE

Fiore, A. T., & Donath, J, S. (2004).Online personals: An overview. *CHI '04 Extended Abstracts on Human Factors in Computing Systems* (pp. 1395–1398). https://doi.org/10.1145/985921.986073

Fitzpatrick, C., & Birnholtz, J. (2018). "I shut the door": Interactions, tensions, and negotiations from a location-based social app. *New Media and Society, 20*(7), 2469–2488. https://doi.org/10.1177/1461444817725064

Fitzpatrick, C, Birnholtz, J., & Brubaker, J. R. (2015). Social and personal disclosure in a location-based real time dating app. *Proceedings of the 48th Annual Hawaii International Conference on System Science* (pp. 1983–1992). https://doi.org/10.1109/HICSS.2015.237

Foucault, M. (1979). *The history of sexuality* (R. Hurley, Trans.). London, UK: Allen Lane.

Fox, J., & McEwan, B. (2017). Distinguishing technologies for social interaction: The perceived social affordances of communication channels scale. *Communication Monographs, 84*(3), 298–318. https://doi.org/10.1080/03637751.2017.1332418

Fredrickson, B. L., & Roberts, T.-A. (1997). Objectification theory: Toward understanding women's lived experiences and mental health risks. *Psychology of Women Quarterly, 21*(2), 173–206. https://doi.org/10.1111/j.1471-6402.1997.tb00108.x

Frizzo-Barker, J., & Chow-White, P. A. (2012). "There's an app for that": Mediating mobile moms and connected careerists through smartphones and networked individualism. *Feminist Media Studies, 12*(4), 580–589. https://doi.org/10.1080/14680777.2012.74187

Fuchs, C. (2016). Baidu, Weibo and Renren: The global political economy of social media in China. *Asian Journal of Communication, 26*(1), 14–41. https://doi.org/10.1080/01292986.2015.1041537

Gaetano, A. (2014). "Leftover women": Postponing marriage and renegotiating womanhood in urban China. *Journal of Research in Gender Studies, 4*(2), 124–149.

Gaetano, A. (2015). *Out to work: Migration, gender, and the changing lives of rural women in contemporary China.* Honolulu, HI: University of Hawai'i Press.

Gibson, J. (1979). *The ecological approach to visual perception.* Boston, MA: Houghton Mifflin.

Giddens, A. (1992). *The transformation of intimacy: Sexuality, love, and eroticism in modern societies.* Stanford, CA: Stanford University Press.

Goffman, E. (1976). *Gender advertisements.* Cambridge, MA: Harvard University Press.

Gluckman, M. (1963). Papers in honor of Melville J. Herskovits: Gossip and scandal. *Current Anthropology, 4*(3), 307–316. https://doi.org/10.1086/200378

Gray, M. L. (2009). *Out in the country: Youth, media, and queer visibility in rural America.* New York, NY: New York University Press.

Green, A. I. (2014). The sexual fields framework. In A. I. Green (Ed.), *Sexual fields: Toward a sociology of collective sexual life* (pp. 25–56). Chicago, IL: University of Chicago Press.

Gross, L. (1993). *Contested closets: The politics and ethics of outing.* Minneapolis, MN: University of Minnesota Press.

Gross, L., & Woods, J. D. (1999). Queers in cyberspace. In L. Gross & J. D. Woods (Eds.), *The Columbia reader on lesbians and gay men in media, society, and politics* (pp. 527–530). New York, NY: Columbia University Press.

Guangdong Bureau of Statistics. (2019). *Guangdong tongji nianjian 2019 nian* 广东统计年鉴 2019年 [Guangdong Statistical Yearbook 2019]. Retrieved from http://stats.gd.gov.cn/gdtjnj/content/post_2639622.html

Guangdong Communication Administration. (2019). *2019 nian 11 yue Guangdong sheng tongxin fazhan qingkuang* 2019年11月广东省通信发展情况 [Communications Development in Guangdong Province in November 2019]. Retrieved from https://gdca.miit.gov.cn/gdcmsnet/gdcms/content/staticView?path=/156/6898.html

Guangzhou changzhu renkou zengliang lingpao guonei yixian chengshi 广州常住人口增量领跑国内一线城市 [Increase in Guangzhou permanent residents surpasses first-tier cities in China]. (2017, March 6). *Nanfang Ribao.* Retrieved from http://www.gz.gov.cn/gzgov/s2342/201703/653121e7cc154b8991f68038f8bf97a4.shtml

Guanyu Momo 关于陌陌 [About Momo]. (2019). Retrieved from https://www.immomo.com/aboutus.html

Gudelunas, D. (2008). *Confidential to America: Newspaper advice columns and sexual education.* New Brunswick, NJ: Transaction.

Gudelunas, D. (2012). There's an app for that: The uses and gratifications of online social networks for gay men. *Sexuality and Culture, 16*(4), 347–365. https://doi.org/10.1007/s12119-012-9127-4

Guest, G., Bunce, A., & Johnson, L. (2006). How many interviews are enough? An experiment with data saturation and variability. *Field Methods, 18*(1), 59–82. https://doi/org/10.1177/1525822x05279903

A guy who grew up in Stockholm's suburbs just sold the "Tinder of China" for $735 million. (2018, February 23). *Business Insider Nordic.* Retrieved from https://nordic.businessinsider.com/a-guy-who-grew-up-in-stockholms-suburbs-just-sold-the-tinder-of-china-for-$735-million--/

Hancock, A.-M. (2016). *Intersectionality: An intellectual history.* New York, NY: Oxford University Press.

Hakim, J. (2018). "The spornosexual": The affective contradictions of male body-work in neoliberal digital culture. *Journal of Gender Studies, 27*(2), 231–241. https://doi.org/10.1080/09589236.2016.1217771

Halberstam, J. (2005). *In a queer time and place: Transgender bodies, subcultural lives.* New York, NY: New York University Press.

Halperin, D. M., & Traub, V. (Eds.). (2009). *Gay shame.* Chicago, IL: University of Chicago Press.

Hanckel, B., Vivienne, S., Byron, P., Robards, B., & Churchill, B. (2019). "That's not necessarily for them": LGBTIQ+ young people, social media platform affordances and identity curation. *Media, Culture and Society, 41*(8), 1261–1278. https://doi.org/10.1177/0163443719846612

Hanisch, C. (2006, January). The personal is political. Retrieved from http://www.carolhanisch.org/CHwritings/PersonalIsPol.pdf

Haraway, D. (1988). Situated knowledges: The science question in feminism and the privilege of partial perspective. *Feminist Studies, 14*(3), 575–599. https://doi.org/10.2307/3178066

Harding, S. (1986). *The science question in feminism.* Ithaca, NY: Cornell University Press.

Harding, S. (1993). Rethinking standpoint epistemology: What is strong objectivity? In L. Alcoff & E. Potter (Eds.), *Feminist epistemologies* (pp. 49–82). New York, NY: Routledge.

Harvey, D. (2005). *A brief history of neoliberalism.* Oxford, UK: Oxford University Press.

Haywood, C. (2018). *Men, masculinity and contemporary dating.* London, UK: Palgrave Macmillan.

Hinsch, B. (1990). *Passions of the cut sleeve: The male homosexual tradition in China.* Berkeley, CA: University of California Press.

Hird, D. (2016). Making class and gender: White-collar men in postsocialist China. In K. Louie (Ed.), *Changing Chinese masculinities: From imperial pillars of state to global real men* (pp. 137–156). Hong Kong, China: Hong Kong University Press.

Hjorth, L. (2003). Pop and *ma*: The landscape of Japanese commodity characters and subjectivity. In C. Berry, F. Martin, & A. Yue (Eds.), *Mobile cultures: New media in queer Asia* (pp. 158–179). Durham, NC: Duke University Press.

Hjorth, L. (2008). *Mobile media in the Asia-Pacific: Gender and the art of being mobile.* Abingdon, UK: Routledge.

Hjorth, L., & Arnold, M. (2013). *Online@AsiaPacific: Mobile, social and locative media in the Asia-Pacific.* London, UK: Routledge.

Ho, L. W. W. (2010). *Gay and lesbian subculture in urban China.* London, UK: Routledge.

Hobbs, M., Owen, S., & Gerber, L. (2017). Liquid love? Dating apps, sex, relationships and the digital transformation of intimacy. *Journal of Sociology, 53*(2), 271–284. https://doi.org/10.1177/1440783316662718

Holter, Ø. G. (2003). *Can men do it? Men and gender equality—The Nordic experience.* Copenhagen, Denmark: Nordic Council of Ministers.

Hong Fincher, L. (2014). *Leftover women: The resurgence of gender inequality in China.* London, UK: Zed Books.

Howson, R. (2006). *Challenging hegemonic masculinity.* London, UK: Routledge.

Hull, T. H. (1990). Recent trends in sex ratios at birth in China. *Population and Development Review, 16*(1), 63–83. https://doi.org/10.2307/1972529

Hutchby, I. (2001). Technologies, texts and affordances. *Sociology, 35*(2), 441–456. https://doi.org/10.1017/S0038038501000219

It is still possible to obtain the exact location of millions of cruising men on Grindr. (2018, September 13). *Queer Europe.* Retrieved from https://www.queereurope.com/it-is-still-possible-to-obtain-the-exact-location-of-cruising-men-on-grindr/

Ito, M. (2008). Introduction. In K. Varneli (Ed.), *Networked publics* (pp. 1–14). Cambridge, MA: MIT Press.

Jin, Y., Manning, K. E., & Chu, L. (2006). Rethinking the "iron girls": Gender and labour during the Chinese Cultural Revolution. *Gender and History, 18*(3), 613–634. https://doi.org/10.1111/j.1468-0424.2006.00458.x

Johansson, S., & Nygren, O. (1991). The missing girls of China: A new demographic account. *Population and Development Review, 17*(1), 35–51. https://doi.org/10.2307/1972351

Johnson, D. G. (2010). Sorting out the question of feminist technology. In L. L. Layne, S. L. Vostral, & K. Boyer (Eds.), *Feminist technology* (pp. 36–54). Urbana, IL: University of Illinois Press.

Jones, R. H. (2005). "You show me yours, I'll show you mine": The negotiation of shifts from textual to visual modes in computer-mediated interaction among gay men. *Visual Communication, 4*(1), 69–92. https://doi.org/10.1177/1470357205048938

Jones, T. (2011, September 22). William Gibson: Beyond cyberspace. *The Guardian.* Retrieved from https://www.theguardian.com/books/2011/sep/22/william-gibson-beyond-cyberspace

Jørgensen, K. M. (2016). The media go-along: Researching mobilities with media at hand. *MedieKultur: Journal of Media and Communication Research, 32*(60), 32–49. https://doi.org/10.7146/mediekultur.v32i60.22429

Judd, E. (2002). *The Chinese women's movement between state and market.* Stanford, CA: Stanford University Press.

Junhong, C. (2001). Prenatal sex determination and sex-selective abortion in rural central China. *Population and Development Review, 27*(2), 259–281. https://doi.org/10.1111/j.1728-4457.2001.00259.x

Kam, L. Y. L. (2013). *Shanghai lalas: Female tongzhi communities and politics in urban China.* Hong Kong, China: Hong Kong University Press.

Kang, W. (2009). *Obsession: Male same-sex relations in China, 1900–1950.* Hong Kong, China: Hong Kong University Press.

Katz, E., Blumler, J. G., & Gurevitch, M. (1973). Uses and gratifications research. *The Public Opinion Quarterly, 37*(4), 509–523. https://doi.org/10.1086/268109

Kennedy, E. L., & Davis, M. (1993). *Boots of leather, slippers of gold: The history of a lesbian community.* New York, NY: Routledge.

Kipnis, A. (2006). Suzhi: A keyword approach. *The China Quarterly, 186*, 295–313. https://doi.org/10.1017/S0305741006000166

Kline, R., & Pinch, T. J. (1996). Users as agents of technological change. *Technology and Culture, 37*(4), 763–795. https://doi.org/10.2307/3107097

Kong, T. S. K. (2011). *Chinese male homosexualities: Memba, tongzhi, and golden boy.* London, UK: Routledge.

Kong, T. S. K., Mahoney, D., & Plummer, K. (2003). Queering the interview. In J. A. Holstein & J. F. Gubrium (Eds.), *Inside interviewing: New lenses, new concerns* (pp. 91–110). Thousand Oaks, CA: Sage.

Kraus, R. (2018, May 31). Grindr, other dating apps are working to add STD notification features. *Mashable*. Retrieved from https://mashable.com/2018/05/31/grindr-tinder-department-of-health-std-notification/

Kuntsmann, A. (2012). Introduction: Affective fabrics of digital cultures. In A. Karatzogianni & A. Kuntsmann (Eds.), *Digital cultures and the politics of emotion: Feelings, affect and technological change*. Basingstoke, UK: Palgrave Macmillan.

Landovitz, R. J., Tseng, C.-H., Weissman, M., Haymer, M., Mendenhall, B., Rogers, K., . . . Shoptaw, S. (2013). Epidemiology, sexual risk behavior, and HIV prevention practices of men who have sex with men using Grindr in Los Angeles, California. *Journal of Urban Health: Bulletin of the New York Academy of Medicine, 90*(4), 729–739. https://doi.org/10.1007/s11524-012-9766-7

Landström, C. (2007). Queering feminist technology studies. *Feminist Studies, 8*(1), 7–26. https://doi.org/10.1177/1464700107074193

Layne, L. L. (2010). Introduction. In L. L. Layne, S. L. Vostral, & K. Boyer (Eds.), *Feminist technology* (pp. 1–35). Urbana, IL: University of Illinois Press.

The L chuangshiren Lu Lei shuo: "Rela shi wo song gei lala pengyoumen de yi ge liwu." The L 创始人鲁磊说: "热拉是我送给拉拉朋友们的一个礼物." [The founder of The L, Lu Lei, said: "Rela is a gift I gave to my lesbian friends"]. (2015, December 28). *Sina News*. Retrieved from http://news.sina.com.cn/o/2015-12-28/doc-ifxmykrf2518851.shtml

Lee, J. (2019). Mediated superficiality and misogyny through cool on Tinder. *Social Media+ Society, 5*(3), 1–11. https://doi.org/10.1177/2056305119872949

Lefebvre, L. E. (2018). Swiping me off my feet: Explicating relationship initiation on Tinder. *Journal of Social and Personal Relationships, 35*(9), 1205–1229. https://doi.org/10.1177/0265407517706419

Levant, R. F., Hall, R. J., & Rankin, T. J. (2013). Male Role Norms Inventory—Short Form (MRNI-SF): Development, confirmatory factor analytic investigation of structure, and measurement invariance across gender. *Journal of Counseling Psychology, 60*(2), 228–238. https://doi.org/10.1037/a0031545

Li, Y. (2014). *Xin zhongguo xing huayu yanjiu 新中国性话语研究* [Studies of sexual discourse in the new China]. Shanghai, China: Shanghai Shehuikexueyuan Chubanshe.

Licoppe, C., Rivière, C. A., & Morel, J. (2016). Grindr casual hook-ups as interactional achievements. *New Media and Society, 18*(11), 2540–2558. https://doi.org/10.1177/1461444815589702

Light, B. (2013). Networked masculinities and social networking sites: A call for the analysis of men and contemporary digital. *Masculinities and Social Change, 2*(3), 245–265. https://doi.org/10.4471/MCS.2013.34

Light, B., Burgess, J., & Duguay, S. (2018). The walkthrough method: An approach to the study of apps. *New Media and Society, 20*(3), 881–900. https://doi.org/10.1177/1461444816675438

Lin, T. (2017, April 28). End of the line for subway ad against sexual harassment. *Sixth Tone.* Retrieved from http://www.sixthtone.com/news/1000123/end-of-the-line-for-subwayad-against-sexual-harassment

Liu, D., & Lu, L. (2005). *Zhongguo tongxinglian yanjiu* 中国同性恋研究 [Studies of Chinese homosexuality]. Beijing, China: Zhongguo Shehui Chubanshe.

Liu, F. (2019). Chinese young men's construction of exemplary masculinity: The hegemony of *chenggong. Men and Masculinities, 22*(2), 294–316. https://doi.org/10.1177/1097184X17696911

Liu, J. (2007). *Gender and work in urban China: Women workers of the unlucky generation.* London, UK: Routledge.

Liu, J.-P., & Ding, N. (2005). Reticent poetics, queer politics. *Inter-Asia Cultural Studies, 6*(1), 30–55. https://doi.org/10.1080/1462394042000326897

Liu, J. X., & Choi, K. (2006). Experiences of social discrimination among men who have sex with men in Shanghai, China. *AIDS and Behavior, 10*(S1), 25–33. https://doi.org/10.1007/s10461-006-9123-5

Liu, T. (2016). Neoliberal ethos, state censorship and sexual culture: A Chinese dating/hook-up app. *Continuum: Journal of Media and Cultural Studies, 30*(5), 557–566. https://doi.org/10.1080/10304312.2016.1210794

Liu, T. (2017). LESDO: Emerging digital infrastructures of community-based care for female queer subjects. *Feminist Media Studies, 17*(2), 301–305. https://doi.org/10.1080/14680777.2017.1283747

Liu, X. (2015). No fats, femmes, or Asians. *Moral Philosophy and Politics, 2*(2), 255–276. https://doi.org/10.1515/mopp-2014-0023

Livingstone, S. (2005). *Audience and publics: When cultural engagement matters for the public sphere.* Portland, OR: Intellect.

Louie, K. (2002). *Theorising Chinese masculinity: Society and gender in China.* Cambridge, UK: Cambridge University Press.

Louie, K. (2015). *Chinese masculinities in a globalizing world.* Abingdon, UK: Routledge.

Lutz, C., & Ranzini, G. (2017). Where dating meets data: Investigating social and institutional privacy concerns on Tinder. *Social Media and Society, 3*(1), 1–12. https://doi.org/10.1177/2056305117697735

MacKenzie, D., & Wajcman, J. (Eds.). (1985). *The social shaping of technology: How the refrigerator got its hum.* Milton Keynes, UK: Open University Press.

Mainwaring, S. D., Chang, M. F., & Anderson, K. (2004). Infrastructures and their discontents: Implications for UbiComp. In N. Davies, E. D. Mynatt, & I. Siio (Eds.), *UbiComp 2004: Ubiquitous Computing* (pp. 418–432). Berlin, Germany: Springer-Verlag. https://doi.org/10.1007/978-3-540-30119-6_25

Marinucci, M. (2010). *Feminism is queer: The intimate connection between queer and feminist theory*. London, UK: Zed Books.

Martin, F. (2010). *Backward glances: Contemporary Chinese cultures and the female homoerotic imaginary*. Durham, NC: Duke University Press.

Martin, P. Y., & Hummer, R. A. (1989). Fraternities and rape on campus. *Gender and Society, 3*(4), 457–473. https://doi.org/10.1177/089124389003004004

Maslow, A. H. (1943). A theory of human motivation. *Psychological Review, 50*(4), 370–396. https://doi.org/10.1037/h0054346

McGlotten, S. (2013). *Virtual intimacies: Media, affect, and queer sociality*. Albany, NY: SUNY Press.

McLelland, M. J. (2002). Virtual ethnography: Using the Internet to study gay culture in Japan. *Sexualities, 5*(4), 387–406. https://doi.org/10.1177/1363460702005004001

McLuhan, M. (1964). *Understanding media: The extensions of human*. New York, NY: McGraw-Hill.

McQuail, D. (2010). *McQuail's mass communication theory* (6th ed.). London, UK: Sage.

Messner, M. A., Greenberg, M. A., & Peretz, T. (2015). *Some men: Feminist allies in the movement to end violence against women*. New York, NY: Oxford University Press.

Miao, W., & Chan, L. S. (2020). Social constructivist account of the world's largest gay social app: Case study of Blued in China. *The Information Society: An International Journal, 36*(4), 214–225. https://doi.org/10.1080/01972243.2020.1762271

Miles, M. B., Huberman, A. M., & Saldaña, J. (2014). *Qualitative data analysis: A methods sourcebook* (3rd ed.). Thousand Oaks, CA: Sage.

Miles, S. (2017). Sex in the digital city: Location-based dating apps and queer urban life. *Gender, Place and Culture, 24*(11), 1595–1610. https://doi.org/10.1080/0966369X.2017.1340874

Miles, S. (2018). Still getting it on online: Thirty years of queer male spaces brokered through digital technologies. *Geography Compass, 12*(11), e12407. https://doi.org/10.1111/gec3.12407

Miller, B. (2015a). "Dude, where's your face?" Self-presentation, self-description, and partner preferences on a social networking application for men who have sex with men: A content analysis. *Sexuality and Culture, 19*(4), 637–658. https://doi.org/10.1007/s12119-015-9283-4

Miller, B. (2015b). "They're the modern-day gay bar": Exploring the uses and gratifications of social networks for men who have sex with men. *Computers in Human Behavior, 51*(Part A), 476–482. https://doi.org/10.1016/j.chb.2015.05.023

Millett, K. (1978). *Sexual politics*. Garden City, NY: Ballantine Books.

Molldrem, S., & Thakor, M. (2017). Genealogies and futures of queer STS: Issues in theory, method, and institutionalization. *Catalyst: Feminism, Theory, Technoscience, 3*(1), 1–15. https://doi.org/10.28968/cftt.v3i1.28795

Molotch, H. (2003). *Where stuff comes from: How toasters, toilets, cars, computers and many other things come to be as they are*. London, UK: Taylor & Francis.

Mowlabocus, S. (2010). *Gaydar culture: Gay men technology and embodiment in the digital age*. Farnham, UK: Ashgate.

Mulvey, L. (1975). Visual pleasure and narrative cinema. *Screen, 16*(3), 6–18. https://doi.org/10.1093/screen/16.3.6

Muñoz, J. E. (2009). *Cruising utopia: The then and there of queer futurity*. New York, NY: New York University Press.

Nagy, P., & Neff, G. (2015). Imagined affordance: Reconstructing a keyword for communication theory. *Social Media and Society, 1*(2), 1–9. https://doi.org/10.1177/2056305115603385

Neff, G., Jordan, T., McVeigh-Schultz, J., & Gillespie, T. (2012). Affordances, technical agency, and the politics of technologies of cultural production. *Journal of Broadcasting and Electronic Media, 56*(2), 299–313. https://doi.org/10.1080/08838151.2012.678520

Newby, J. (2018, June 8). "We will not give up": China's biggest pride event turns 10. *RADII*. Retrieved from https://radiichina.com/we-will-not-give-up-chinas-biggest-pride-event-turns-10/

Newton, E. (1993). My best informant's dress: The erotic equation in fieldwork. *Cultural Anthropology, 8*(1), 3–23.

Norma, C. (2011). Prostitution and the 1960s' origins of corporate entertaining in Japan. *Women's Studies International Forum, 34*(6), 509–519. https://doi.org/10.1016/j.wsif.2011.07.005

Norman, D. A. (1988). *The psychology of everyday things*. New York, NY: Basic Books.

Oldenburg, R., & Brissett, D. (1982). The third place. *Qualitative Sociology, 5*(4), 265–284. https://doi.org/doi:10.1007/BF00986754

O'Neill, R. (2018). *Seduction: Men, masculinity, and mediated intimacy*. Medford. MA: Polity.

Online outcry forces China's Twitter, Weibo, to backtrack on censorship of gay content. (2018, April 16). *National Public Radio*. Retrieved from https://www.scmp

.com/news/china/society/article/2141907/online-outcry-forces-chinas-twitter
-sina-weibo-backtrack

Oudshoorn, N., & Pinch, T. (Eds.). (2003). *How users matter: The co-construction of users and technologies.* Cambridge, MA: MIT Press.

Paasonen, S., Light, B., & Jarrett, K. (2019). The dick pic: Harassment, curation, and desire. *Social Media and Society, 5*(2), 1–10. https://doi.org/10.1177/2056305119826126

Padilla, M., Hirsch, J., Muñoz-Laboy, M., Sember, R., & Parker, R. (Eds.). (2007). *Love and globalization: Transformations of intimacy in the contemporary world.* Nashville, TN: Vanderbilt University Press.

Papacharissi, Z. (2014). *Affective publics: Sentiment, technology, and politics.* New York, NY: Oxford University Press.

Parry, L. (2015, May 27). Tinder and Grindr dating apps "are causing cases of syphilis, gonorrhoea and HIV to soar," experts claim. *Daily Mail.* Retrieved from https://www.dailymail.co.uk/health/article-3098849/Tinder-Grindr-dating-apps-causing-cases-syphilis-gonorrhoea-HIV-soar-experts-claim.html

Pei, Y. (2013). *Yuwang dushi: Shanghai 70 hou nüxing yanjiu* 欲望都市: 上海70后女性研究 [Sex and the city: Studies of post-70s Shanghainese women]. Shanghai, China: Shanghai Renmin Chubanshe.

Pew Research Center. (2013, June 4). The global divide on homosexuality. Retrieved from https://www.pewglobal.org/2013/06/04/the-global-divide-on-homosexuality/

Pinch, T. J., & Bijker, W. E. (1987). The social construction of facts and artifacts: Or how the sociology of science and the sociology of technology might benefit each other. In W. E. Bijker, T. P. Hughes, & T. J. Pinch (Eds.), *The social construction of technological systems: New directions in the sociology and history of technology* (pp. 17–50). Cambridge, MA: MIT Press.

Plant, S. (1997). *Zeros + ones: Digital women + the new technoculture.* New York, NY: Doubleday.

Plutchik, R. (1979). *Emotion: A psychoevolutionary synthesis.* New York, NY: Harper & Row.

Portwood-Stacer, L. (2013). Media refusal and conspicuous non-consumption: The performative and political dimensions of Facebook abstention. *New Media and Society, 15*(7), 1041–1057. https://doi.org/10.1177/1461444812465139

Poster, J. (2002). Trouble, pleasure, and tactics: Anonymity and identity in a lesbian chat room. In M. Consalvo & S. Paasonen (Eds.), *Women and everyday uses of the internet: Agency and identity* (pp. 230–252). New York, NY: Peter Lang.

Putnam, R. D. (2000). *Bowling alone: The collapse and revival of American community.* New York, NY: Simon & Schuster.

Qiu, J. L. (2009). *Working-class network society: Communication technology and the information have-less in urban China*. Cambridge, MA: MIT Press.

Qiu, Z. (2013). Cuteness as a subtle strategy: Urban female youth and the online *feizhuliu* culture in contemporary China. *Cultural Studies, 27*(2), 225–241. https://doi.org/10.1080/09502386.2012.738640

Race, K. (2015). "Party and play": Online hook-up devices and the emergence of PNP practices among gay men. *Sexualities, 18*(3), 253–275. https://doi.org/10.1177/1363460714550913

Ragin, C. C., & Fiss, P. C. (2017). *Intersectional inequality: Race, class, test scores, and poverty*. Chicago, IL: University of Chicago Press.

Rainie, H., & Wellman, B. (2012). *Networked: The new social operating system*. Cambridge, MA: MIT Press.

Rand, E. J. (2012). Gay pride and its queer discontents: ACT UP and the political deployment of affect. *Quarterly Journal of Speech, 98*(1), 75–80. https://doi.org/10.1080/00335630.2011.638665

Rauniar, R., Rawski, G., Yang, J., & Johnson, B. (2014). Technology acceptance model (TAM) and social media usage: An empirical study on Facebook. *Journal of Enterprise Information Management, 27*(1), 6–30. https://doi.org/10.1108/JEIM-04-2012-0011

Reinharz, S., & Chase, S. E. (2003). Interviewing women. In J. A. Holstein & J. F. Gubrium (Eds.), *Inside interviewing: New lenses, new concerns* (pp. 73–90). Thousand Oaks, CA: Sage.

Reynolds, D. (2018, December 3). Tumblr's ban on adult content alarms LGBTQ Twitter. *Advocate*. Retrieved from https://www.advocate.com/business/2018/12/03/tumblrs-ban-adult-content-alarms-lgbtq-twitter

Rice, E., Holloway, I., Winetrobe, H., Rhoades, H., Barman-Adhikari, A., Gibbs, J., . . . Dunlap, S. (2012). Sex risk among young men who have sex with men who use Grindr, a smartphone geosocial networking application. *AIDS and Clinical Research, S4*(5), 1–8. https://doi.org/10.4172/2155-6113.S4-005

Rich, A. (1993). Compulsory heterosexuality and lesbian existence. In H. Abelove, M. A. Barale, & D. M. Halperin (Eds.), *The lesbian and gay studies reader* (pp. 227–254). London, UK: Routledge.

Rochadiat, A. M. P., Tong, S. T., & Novak, J. M. (2018). Online dating and courtship among Muslim American women: Negotiating technology, religious identity, and culture. *New Media and Society, 20*(4), 1618–1639. https://doi.org/10.1177/1461444817702396

Rofel, L. (2007). *Desiring China: Experiments in neoliberalism, sexuality, and public culture*. Durham, NC: Duke University Press.

Rongzi baiwan meijin, liang ge da nansheng ruhe dazao lala shequ 融资百万美金, 两个大男生如何打造拉拉社区 [Raising USD1 million, how did two big boys create a lesbian community]. (2016, January 7). *Yiwang Keji*. Retrieved from http://tech.163 .com/16/0107/09/BCNE2FHH00094OE0.html

Roth, Y. (2014). Locating the "Scruff guy": Theorizing body and space in gay geosocial media. *International Journal of Communication, 8*, 2113–2133.

Rothblum, E. D. (2010). The complexity of butch and femme among sexual minority women in the 21st century. *Psychology of Sexualities Review, 1*(1), 29–42.

Rubin, G. (1984). Thinking sex: Notes for a radical theory of the politics of sexuality. In C. S. Vance (Ed.), *Pleasure and danger: Exploring female sexuality* (pp. 267–319). Boston, MA: Routledge & Paul.

Sang, T. D. (2003). *The emerging lesbian: Female same-sex desire in modern China*. Chicago, IL: University of Chicago Press.

Sanger, D. E. (2019, March 28). Grindr is owned by a Chinese firm, and the U.S. is trying to force it to sell. *The New York Times*. Retrieved from https://www.nytimes .com/2019/03/28/us/politics/grindr-china-national-security.html

Sawyer, A. N., Smith, E. R., & Benotsch, E. G. (2018). Dating application use and sexual risk behavior among young adults. *Sexuality Research and Social Policy, 15*(2), 183–191. https://doi.org/10.1007/s13178-017-0297-6

Scarantino, A. (2003). Affordances explained. *Philosophy of Science, 70*(5), 949–961. https://doi.org/10.1086/377380

Schaefer, D. O. (2015). *Religious affects: Animality, evolution, and power*. Durham, NC: Duke University Press.

Schrock, A. R. (2015). Communicative affordances of mobile media: Portability, availability, locatability, and multimediality. *International Journal of Communication, 9*, 1229–1246.

Schrock, D., & Schwalbe, M. (2009). Men, masculinity, and manhood acts. *Annual Review of Sociology, 35*, 277–295. https://doi.org/10.1146/annurev-soc-070308-115933

Schwalbe, M. L., & Wolkomir, D. (2003). Interviewing men. In J. A. Holstein & J. F. Gubrium (Eds.), *Inside interviewing: New lenses, new concerns* (pp. 55–72). Thousand Oaks, CA: Sage.

Sedgwick, E. K. (1993). Queer performativity: Henry James's *The Art of the Novel*. *GLQ: A Journal of Lesbian and Gay Studies, 1*(1), 1–16. https://doi.org/10.1215/10642684 -1-1-1

Sedgwick, E. K. (2003). *Touching feeling: Affect, pedagogy, performativity*. Durham, NC: Duke University Press.

Sedgwick, E. K., & Frank, A. (Eds.). (1995). *Shame and its sisters: A Silvan Tomkins reader*. Durham, NC: Duke University Press.

Sender, K. (2017). Expanding media and sexuality studies: A transnational study of sex museums. *Critical Studies in Media Communication, 34*(1), 73–79. https://doi.org/ 10.1080/15295036.2016.1266685

Shen, Y. (2011). China in the "post-patriarchal era": Changes in the power relationships in urban households and an analysis of the course of gender inequality in society. *Chinese Sociology & Anthropology, 43*(4), 5–23. https://doi.org/10.2753/ CSA0009-4625430401

Shi, L. (2014). *Chinese lesbian cinema: Mirror rubbing, lala, and les.* Lanham, MD: Lexington Books.

Shouse, E. (2005). Feeling, emotion, affect. *M/C Journal, 8*(6). Retrieved from http:// journal.media-culture.org.au/0512/03-shouse.php

Siebler, K. (2016). *Learning queer identity in the digital age.* New York, NY: Palgrave Macmillan.

Simpson, M. (1999). *It's a queer world: Deviant adventures in pop culture.* New York, NY: Harrington Park Press.

Solis, R. J. C., & Wong, K. Y. J. (2019). To meet or not to meet? Measuring motivations and risks as predictors of outcomes in the use of mobile dating applications in China. *Chinese Journal of Communication, 12*(2), 204–223. https://doi.org/10.1080/ 17544750.2018.1498006

Song, D. (2016). *Jindai lingnan wenhua jiazhiguan de yanbian* 近代岭南文化价值观的演变 [The evolution of modern Lingnan cultural values]. Guangzhou, China: Sun Yat-Sen University Press.

Suler, J. (2004). The online disinhibition effect. *Cyberpsychology and Behavior, 7*(3), 321–326. https://doi.org/10.1089/1094931041291295

Sumter, S. R., & Vandenbosch, L. (2019). Dating gone mobile: Demographic and personality-based correlates of using smartphone-based dating applications among emerging adults. *New Media and Society, 21*(3), 655–673. https://doi.org/10.1177/ 1461444818804773

Sumter, S. R., Vandenbosch, L., & Ligtenberg, L. (2017). Love me Tinder: Untangling emerging adults' motivations for using the dating application Tinder. *Telematics and Informatics, 34*(1), 67–78. https://doi.org/10.1016/j.tele.2016.04.009

Tait, A. (2017, August 30). Swipe right for equality: How Bumble is taking on sexism. *Wired.* Retrieved from https://www.wired.co.uk/article/bumble-whitney-wolfe-sexism -tinder-app

Tan, J. (2017). Digital masquerading: Feminist media activism in China. *Crime, Media, Culture, 13*(2), 171–186. https://doi.org/10.1177/1741659017710063

Tan, J. (2018, April). *Gendering the platforms: Women's social app videos.* Paper presented at The Platformization of Chinese Society: An International Workshop, Hong Kong, China.

Tang, D. T.-S. (2015). Essential labels? Gender identity politics on Hong Kong mobile phone application. In L. Hjorth & O. Khoo (Eds.), *Routledge handbook of new media in Asia* (pp. 263–274). London, UK: Routledge.

Tang, D. T.-S. (2017). All I get is an emoji: Dating on lesbian mobile phone app Butterfly. *Media, Culture and Society, 39*(6), 816–832. https://doi.org/10.1177/0163443717693680

Tang, Y. (2014, February 9). Ruhe pingjia Momo yonghu guo yi? 如何评价陌陌用户过亿? [How to comment on Momo's users exceeding 100 million?] [Web log post]. Retrieved from https://www.zhihu.com/question/22680523

Timmermans, E., & Courtois, C. (2018). From swiping to casual sex and/or committed relationships: Exploring the experiences of Tinder users. *The Information Society: An International Journal, 34*(2), 59–70. https://doi.org/10.1080/01972243.2017.1414093

Timmermans, E., & De Caluwé, E. (2017a). Development and validation of the Tinder Motives Scale (TMS). *Computers in Human Behavior, 70*, 341–350. https://doi.org/10.1016/j.chb.2017.01.028

Timmermans, E., & De Caluwé, E. (2017b). To Tinder or not to Tinder, that's the question: An individual differences perspective to Tinder use and motives. *Personality and Individual Differences, 110*(1), 74–79. https://doi.org/10.1016/j.paid.2017.01.026

Timmermans, E., De Caluwé, E., & Alexopoulos, C. (2018). Why are you cheating on tinder? Exploring users' motives and (dark) personality traits. *Computers in Human Behavior, 89*, 129–139. https://doi.org/10.1016/j.chb.2018.07.040

To, S. (2015). *China's leftover women: Late marriage among professional women and its consequences*. London, UK: Routledge.

Tomkins, S. S. (1962). *Affect imagery consciousness: Vol. 1. The positive affects*. New York, NY: Springer.

Tomkins, S. S. (1963). *Affect imagery consciousness: Vol. 2. The negative affects*. New York, NY: Springer.

Tong, S. T., Hancock, J. T., & Slatcher, R. B. (2016). Online dating system design and relational decision making: Choice, algorithms, and control. *Personal Relationships, 23*(4), 645–662. https://doi.org/10.1111/pere.12158

2018 nian Zhongguo aizibing ganran renshu, fabing renshu, siwang renshu he chuanbo tujing tongji qingkuang 2018 年中国艾滋病感染人数, 发病人数, 死亡人数和传播途径统计情况 [Statistics on the number of HIV infections, the number of cases, the number of deaths, and the routes of transmission in China 2018]. (2018, November 27). *Zhonggou Baogao Wang*. Retrieved from http://free.chinabaogao.com/yiyao/201811/112ISMR018.html

Van De Wiele, C., & Tong, S. T. (2014). Breaking boundaries: The uses and gratifications of Grindr. *Proceedings of the 2014 ACM International Joint Conference on Pervasive and Ubiquitous Computing* (pp. 619–630). https://doi.org/10.1145/2632048.2636070

Van Gulik, R. (1961). *Sexual life in ancient China: A preliminary survey of Chinese sex and society from ca. 1500 B.C. till 1644 A.D.* Leiden, The Netherlands: E. J. Brill.

Vannewkirk, R. (2006). "Gee, I didn't get that vibe from you": Articulating my own version of a femme lesbian existence. *Journal of Lesbian Studies, 10*(1–2), 73–85. https://doi.org/10.1300/J155v10n01_04

Van Oost, E. (2003). Materialized gender: How shavers configure the users' femininity and masculinity. In N. Oudshoorn & T. Pinch (Eds.), *How users matter: The co-construction of users and technologies* (pp. 193–208). Cambridge, MA: MIT Press.

Vörös, F. (2015). Troubling complicity: Audience ethnography, male porn viewers and feminist critique. *Porn Studies, 2*(2–3), 137–149. https://doi.org/10.1080/23268743.2015.1052936

Wade, L. (2017). *American hookup: The new culture of sex on campus.* New York, NY: W.W. Norton.

Wajcman, J. (1991). *Feminism confronts technology.* University Park, PA: Pennsylvania State University Press.

Wajcman, J. (2006). Technocapitalism meets technofeminism. *Labour and Industry, 16*(3), 7–20. https://doi.org/10.1080/10301763.2006.10669327

Wajcman, J. (2007). From women and technology to gendered technoscience. *Information, Communication and Society, 10*(3), 287–298. https://doi.org/10.1080/13691180701409770

Wallis, C. (2013). *Technomobility in China: Young migrant women and mobile phones.* New York, NY: New York University Press.

Wang, S. (2019a). Chinese affective platform economies: Dating, live streaming, and performative labor on Blued. *Media, Culture & Society.* https://doi.org/10.1177/0163443719867283

Wang, S. (2019b). Live streaming, intimate situations, and the circulation of same-sex affect: Monetizing affective encounters on Blued. *Sexualities.* https://doi.org/10.1177/1363460719872724

Wang, S. (2020). Calculating dating goals: Data gaming and algorithmic sociality on Blued, a Chinese gay dating app. *Information, Communication and Society, 23*(2), 181–197. https://doi.org/10.1080/1369118X.2018.1490796

Wang, Z. (2005). "State feminism"? Gender and socialist state formation in Maoist China. *Feminist Studies, 31*(3), 519–551. https://doi.org/10.2307/20459044

Ward, J. (2017). What are you doing on Tinder? Impression management on a matchmaking mobile app. *Information, Communication and Society, 20*(11), 1644–1659. https://doi.org/10.1080/1369118X.2016.1252412

Warf, B. (2006). Infrastructure. In *Encyclopedia of Human Geography* (p. 258). Thousand Oaks, CA: Sage.

Wen, C. (2015). The advertising and profit model of leading dating sites in China: A comparison of Jianyuan, Baihe and Zhenai's targeting and advertising. In I. A. Degim, J. Johnson, & T. Fu (Eds.), *Online courtship: Interpersonal interactions across borders* (pp. 106–116). Amsterdam, The Netherlands: Institute of Network Cultures.

West, I., Frischherz, M., Panther, A., & Brophy, R. (2013). Queer worldmaking in the "It gets better" campaign. *QED: A Journal in GLBTQ Worldmaking, Inaugural Issue,* 49–86. https://doi.org/10.14321/qed.0049

White, T. (1994). The origins of China's birth planning policy. In C. Glimartin, G. Herschatter, L. Rofel, & T. White (Eds.), *Engendering China: Women, culture and the state* (pp. 250–278). Cambridge, MA: Harvard University Press.

Whitty, M. (2008). Revealing the "real" me, searching for the "actual" you: Presentations of self on an internet dating site. *Computers in Human Behavior, 24*(4), 1707–1723. https://doi.org/10.1016/j.chb.2007.07.002

Williams, R. (1961). *The long revolution.* London, UK: Chatto & Windus.

Wilson, A. (2016). The infrastructure of intimacy. *Signs: Journal of Women in Culture and Society, 41*(2), 247–280. https://doi.org/10.1086/682919

Winner, L. (1980). Do artifacts have politics? *Daedalus, 109*(1), 121–136.

Wong, A. (2019, May 8). Singapore makes it illegal to send unwanted nudes. *Inkstone.* Retrieved from https://www.inkstonenews.com/society/singapore-makes-it -illegal-send-unwanted-nudes/article/3009336

Wong, D. (2015). Sexual minorities in China. In J. D. Wright (Ed.), *International encyclopedia of social and behavioral sciences* (2nd ed.) (pp. 734–739). Amsterdam, The Netherlands: Elsevier.

Wong, E. (2010, July 26). Move to limit Cantonese on Chinese TV is assailed. *New York Times.* Retrieved from https://www.nytimes.com/2010/07/27/world/asia/ 27cantonese.html

Xing, Q. (2011, December 24). Guangzhou shi changzhu renkou chao 1200 wan 广州市常住人口超1200万 [Guangzhou's resident population exceeds 12 million]. *Ifeng.* Retrieved from http://news.ifeng.com/gundong/detail_2011_12/24/11542983 _0.shtml

Yan, Y. (2009). *The individualization of Chinese society.* Oxford, UK: Berg.

Yang, G. (2017). The online translation activism of bridge bloggers, feminists, and cyber nationalists in China. In V. Pickard & G. Yang (Eds.), *Media activism in the digital age* (pp. 62–75). London, UK: Routledge.

Yang, J. (2011). *Nennu* and *shunu*: Gender, body politics, and the beauty economy in China. *Signs: Journal of Women in Culture and Society, 36*(2), 333–357. https://doi.org/10.1086/655913

Yeo, T. E. D., & Fung, T. H. (2018). "Mr Right Now": Temporality of relationship formation on gay mobile dating apps. *Mobile Media and Communication, 6*(1), 3–18. https://doi.org/10.1177/2050157917718601

Yeo, T. E. D., & Ng, Y. L. (2016). Sexual risk behaviors among apps-using young men who have sex with men in Hong Kong. *AIDS Care, 28*(3), 314–318. https://doi.org/10.1080/09540121.2015.1093597

Zeng, J. (2015, November 6). Zhongguo nüxing zhong qi yisheng chengshou de baoli 中國女性終其一生承受的暴力 [The violence suffered by Chinese women in their whole life]. *The Initium*. Retrieved from https://theinitium.com/article/20151106-opinion-Chinese-female/

Zhang, J., & Sun, P. (2014). "When are you going to get married?" Parental matchmaking and middle-class women in contemporary urban China. In D. Davis & S. Friedman (Eds.), *Wives, husbands, and lovers: Marriage and sexuality in Hong Kong, Taiwan, and urban China* (pp. 118–144). Stanford, CA: Stanford University Press.

Zhang, Y., & Erni, J. (2018). In with expectations and out with disappointment: Gay-tailored social media and the redefinition of intimacy. In R. Andreassen, M. Nebeling, K. Harrison, & T. Raun (Eds.), *Mediated intimacies: Connectivities, relationalities and proximities* (pp. 143–155). Abingdon, UK: Routledge.

Zhang, Y., Tang, L. S.-T., & Leung, L. (2011). Gratifications, collective self-esteem, online emotional openness, and traitlike communication apprehension as predictors of Facebook uses. *Cyberpsychology, Behavior, and Social Networking, 14*(12), 733–739. https://doi.org/10.1089/cyber.2010.0042

Zheng, T. (2015). *Tongzhi living: Men attracted to men in postsocialist China*. Minneapolis, MN: University of Minnesota Press.

Zhou, J. (2006). Chinese prostitution: Consequences and solutions in the post-Mao era. *China: An International Journal, 4*(2), 238–262. https://doi.org/10.1142/S0219747206000136

INDEX

www.ingramcontent.com/pod-product-compliance
Lightning Source LLC
Chambersburg PA
CBHW070331270326
41926CB00017B/3842